Unconscious Memory

Samuel Butler

Contents

UNCONSCIOUS MEMORY

BY

Samuel Butler

UNCONSCIOUS MEMORY

"As this paper contains nothing which deserves the name either of experiment or discovery, and as it is, in fact, destitute of every species of merit, we should have allowed it to pass among the multitude of those articles which must always find their way into the collections of a society which is pledged to publish two or three volumes every year. . . . We wish to raise our feeble voice against innovations, that can have no other effect than to check the progress of science, and renew all those wild phantoms of the imagination which Bacon and Newton put to flight from her temple."--Opening Paragraph of a Review of Dr. Young's Bakerian Lecture. Edinburgh Review, January 1803, p. 450.

"Young's work was laid before the Royal society, and was made the 1801 Bakerian Lecture. But he was before his time. The second number of the Edinburgh Review contained an article levelled against him by Henry (afterwards Lord) Brougham, and this was so severe an attack that Young's ideas were absolutely quenched for fifteen years. Brougham was then only twenty-four years of age. Young's theory was reproduced in France by Fresnel. In our days it is the accepted theory, and is found to explain all the phenomena of light."--Times Report of a Lecture by Professor Tyndall on Light, April 27, 1880.

This Book
Is inscribed to
RICHARD GARNETT, ESQ.
(Of the British Museum)
In grateful acknowledgment of the unwearying kindness with which he
has so often placed at my disposal his varied store of information.

NOTE

For many years a link in the chain of Samuel Butler's biological
works has been missing. "Unconscious Memory" was originally
published thirty years ago, but for fully half that period it has
been out of print, owing to the destruction of a large number of the
unbound sheets in a fire at the premises of the printers some years
ago. The present reprint comes, I think, at a peculiarly fortunate
moment, since the attention of the general public has of late been
drawn to Butler's biological theories in a marked manner by several
distinguished men of science, notably by Dr. Francis Darwin, who, in
his presidential address to the British Association in 1908, quoted
from the translation of Hering's address on "Memory as a Universal
Function of Original Matter," which Butler incorporated into
"Unconscious Memory," and spoke in the highest terms of Butler
himself. It is not necessary for me to do more than refer to the
changed attitude of scientific authorities with regard to Butler and
his theories, since Professor Marcus Hartog has most kindly consented
to contribute an introduction to the present edition of "Unconscious
Memory," summarising Butler's views upon biology, and defining his
position in the world of science. A word must be said as to the

controversy between Butler and Darwin, with which Chapter IV is concerned. I have been told that in reissuing the book at all I am committing a grievous error of taste, that the world is no longer interested in these "old, unhappy far-off things and battles long ago," and that Butler himself, by refraining from republishing "Unconscious Memory," tacitly admitted that he wished the controversy to be consigned to oblivion. This last suggestion, at any rate, has no foundation in fact. Butler desired nothing less than that his vindication of himself against what he considered unfair treatment should be forgotten. He would have republished "Unconscious Memory" himself, had not the latter years of his life been devoted to all-engrossing work in other fields. In issuing the present edition I am fulfilling a wish that he expressed to me shortly before his death.

R. A. STREATFEILD.
April, 1910.

INTRODUCTION By Marcus Hartog, M.A. D.Sc., F.L.S., F.R.H.S.

In reviewing Samuel Butler's works, "Unconscious Memory" gives us an invaluable lead; for it tells us (Chaps. II, III) how the author came to write the Book of the Machines in "Erewhon" (1872), with its foreshadowing of the later theory, "Life and Habit," (1878), "Evolution, Old and New" (1879), as well as "Unconscious Memory" (1880) itself. His fourth book on biological theory was "Luck? or Cunning?" (1887). {0a}

Besides these books, his contributions to biology comprise several essays: "Remarks on Romanes' Mental Evolution in Animals, contained in "Selections from Previous Works" (1884) incorporated into "Luck? or Cunning," "The Deadlock in Darwinism" (Universal Review, April-June, 1890), republished in the posthumous volume of "Essays on Life, Art, and Science" (1904), and, finally, some of the "Extracts from the Notebooks of the late Samuel Butler," edited by Mr. H. Festing Jones, now in course of publication in the New Quarterly Review.

Of all these, "LIFE AND HABIT" (1878) is the most important, the main building to which the other writings are buttresses or, at most, annexes. Its teaching has been summarised in "Unconscious Memory" in four main principles: "(1) the oneness of personality between parent and offspring; (2) memory on the part of the offspring of certain actions which it did when in the persons of its forefathers; (3) the latency of that memory until it is rekindled by a recurrence of the associated ideas; (4) the unconsciousness with which habitual actions come to be performed." To these we must add a fifth: the purposiveness of the actions of living beings, as of the machines which they make or select.

Butler tells ("Life and Habit," p. 33) that he sometimes hoped "that this book would be regarded as a valuable adjunct to Darwinism." He was bitterly disappointed in the event, for the book, as a whole, was received by professional biologists as a gigantic joke--a joke, moreover, not in the best possible taste. True, its central ideas, largely those of Lamarck, had been presented by Hering in 1870 (as Butler found shortly after his publication); they had been favourably received, developed by Haeckel, expounded and praised by Ray Lankester. Coming from Butler, they met with contumely, even from such men as Romanes, who, as Butler had no difficulty in proving, were unconsciously inspired by the same ideas--"Nur mit ein bischen ander'n Worter."

It is easy, looking back, to see why "Life and Habit" so missed its
mark. Charles Darwin's presentation of the evolution theory had, for
the first time, rendered it possible for a "sound naturalist" to
accept the doctrine of common descent with divergence; and so given a
real meaning to the term "natural relationship," which had forced
itself upon the older naturalists, despite their belief in special
and independent creations. The immediate aim of the naturalists of
the day was now to fill up the gaps in their knowledge, so as to
strengthen the fabric of a unified biology. For this purpose they
found their actual scientific equipment so inadequate that they were
fully occupied in inventing fresh technique, and working therewith at
facts--save a few critics, such as St. George Mivart, who was
regarded as negligible, since he evidently held a brief for a party
standing outside the scientific world.

Butler introduced himself as what we now call "The Man in the
Street," far too bare of scientific clothing to satisfy the Mrs.
Grundy of the domain: lacking all recognised tools of science and
all sense of the difficulties in his way, he proceeded to tackle the
problems of science with little save the deft pen of the literary
expert in his hand. His very failure to appreciate the difficulties
gave greater power to his work--much as Tartarin of Tarascon ascended
the Jungfrau and faced successfully all dangers of Alpine travel, so
long as he believed them to be the mere "blagues de reclame" of the
wily Swiss host. His brilliant qualities of style and irony
themselves told heavily against him. Was he not already known for
having written the most trenchant satire that had appeared since
"Gulliver's Travels"? Had he not sneered therein at the very
foundations of society, and followed up its success by a pseudo-
biography that had taken in the "Record" and the "Rock"? In "Life
and Habit," at the very start, he goes out of his way to heap scorn

at the respected names of Marcus Aurelius, Lord Bacon, Goethe, Arnold of Rugby, and Dr. W. B. Carpenter. He expressed the lowest opinion of the Fellows of the Royal Society. To him the professional man of science, with self-conscious knowledge for his ideal and aim, was a medicine-man, priest, augur--useful, perhaps, in his way, but to be carefully watched by all who value freedom of thought and person, lest with opportunity he develop into a persecutor of the worst type. Not content with blackguarding the audience to whom his work should most appeal, he went on to depreciate that work itself and its author in his finest vein of irony. Having argued that our best and highest knowledge is that of whose possession we are most ignorant, he proceeds: "Above all, let no unwary reader do me the injustice of believing in me. In that I write at all I am among the damned."

His writing of "EVOLUTION, OLD AND NEW" (1879) was due to his conviction that scant justice had been done by Charles Darwin and Alfred Wallace and their admirers to the pioneering work of Buffon, Erasmus Darwin, and Lamarck. To repair this he gives a brilliant exposition of what seemed to him the most valuable portion of their teachings on evolution. His analysis of Buffon's true meaning, veiled by the reticences due to the conditions under which he wrote, is as masterly as the English in which he develops it. His sense of wounded justice explains the vigorous polemic which here, as in all his later writings, he carries to the extreme.

As a matter of fact, he never realised Charles Darwin's utter lack of sympathetic understanding of the work of his French precursors, let alone his own grandfather, Erasmus. Yet this practical ignorance, which to Butler was so strange as to transcend belief, was altogether genuine, and easy to realise when we recall the position of Natural Science in the early thirties in Darwin's student days at Cambridge, and for a decade or two later. Catastropharianism was the tenet of the day: to the last it commended itself to his Professors of Botany and Geology,--for whom Darwin held the fervent allegiance of the Indian scholar, or chela, to his guru. As Geikie has recently pointed out, it was only later, when Lyell had shown that the breaks in the succession of the rocks were only partial and local, without involving the universal catastrophes that destroyed all life and rendered fresh creations thereof necessary, that any general acceptance of a descent theory could be expected. We may be very sure that Darwin must have received many solemn warnings against the dangerous speculations of the "French Revolutionary School." He himself was far too busy at the time with the reception and assimilation of new facts to be awake to the deeper interest of far-reaching theories.

It is the more unfortunate that Butler's lack of appreciation on these points should have led to the enormous proportion of bitter personal controversy that we find in the remainder of his biological writings. Possibly, as suggested by George Bernard Shaw, his acquaintance and admirer, he was also swayed by philosophical resentment at that banishment of mind from the organic universe, which was generally thought to have been achieved by Charles Darwin's theory. Still, we must remember that this mindless view is not implicit in Charles Darwin's presentment of his own theory, nor was it accepted by him as it has been by so many of his professed disciples.

"UNCONSCIOUS MEMORY" (1880).--We have already alluded to an anticipation of Butler's main theses. In 1870 Dr. Ewald Hering, one of the most eminent physiologists of the day, Professor at Vienna, gave an Inaugural Address to the Imperial Royal Academy of Sciences: "Das Gedachtniss als allgemeine Funktion der organisirter Substanz" ("Memory as a Universal Function of Organised Matter"). When "Life and Habit" was well advanced, Francis Darwin, at the time a frequent visitor, called Butler's attention to this essay, which he himself only knew from an article in "Nature." Herein Professor E. Ray Lankester had referred to it with admiring sympathy in connection with its further development by Haeckel in a pamphlet entitled "Die Perigenese der Plastidule." We may note, however, that in his collected Essays, "The Advancement of Science" (1890), Sir Ray Lankester, while including this Essay, inserts on the blank page {0b}--we had almost written "the white sheet"--at the back of it an apology for having ever advocated the possibility of the transmission of acquired characters.

"Unconscious Memory" was largely written to show the relation of Butler's views to Hering's, and contains an exquisitely written translation of the Address. Hering does, indeed, anticipate Butler, and that in language far more suitable to the persuasion of the scientific public. It contains a subsidiary hypothesis that memory has for its mechanism special vibrations of the protoplasm, and the acquired capacity to respond to such vibrations once felt upon their repetition. I do not think that the theory gains anything by the introduction of this even as a mere formal hypothesis; and there is no evidence for its being anything more. Butler, however, gives it a warm, nay, enthusiastic, reception in Chapter V (Introduction to Professor Hering's lecture), and in his notes to the translation of the Address, which bulks so large in this book, but points out that he was "not committed to this hypothesis, though inclined to accept it on a prima facie view." Later on, as we shall see, he attached more importance to it.

The Hering Address is followed in "Unconscious Memory" by translations of selected passages from Von Hartmann's "Philosophy of the Unconscious," and annotations to explain the difference from this personification of "The Unconscious" as a mighty all-ruling, all-creating personality, and his own scientific recognition of the great part played by UNCONSCIOUS PROCESSES in the region of mind and memory.

These are the essentials of the book as a contribution to biological philosophy. The closing chapters contain a lucid statement of objections to his theory as they might be put by a rigid necessitarian, and a refutation of that interpretation as applied to human action.

But in the second chapter Butler states his recession from the strong logical position he had hitherto developed in his writings from "Erewhon" onwards; so far he had not only distinguished the living from the non-living, but distinguished among the latter MACHINES or TOOLS from THINGS AT LARGE. {0c} Machines or tools are the external organs of living beings, as organs are their internal machines: they are fashioned, assembled, or selected by the beings for a purposes so they have a FUTURE PURPOSE, as well as a PAST HISTORY. "Things at large" have a past history, but no purpose (so long as some being does not convert them into tools and give them a purpose): Machines have a Why? as well as a How?: "things at large" have a How? only.

In "Unconscious Memory" the allurements of unitary or monistic views have gained the upper hand, and Butler writes (p. 23):-

"The only thing of which I am sure is, that the distinction between the organic and inorganic is arbitrary; that it is more coherent with our other ideas, and therefore more acceptable, to start with every molecule as a living thing, and then deduce death as the breaking up of an association or corporation, than to start with inanimate molecules and smuggle life into them; and that, therefore, what we call the inorganic world must be regarded as up to a certain point living, and instinct, within certain limits, with consciousness, volition, and power of concerted action. IT IS ONLY OF LATE, HOWEVER, THAT I HAVE COME TO THIS OPINION."

I have italicised the last sentence, to show that Butler was more or less conscious of its irreconcilability with much of his most characteristic doctrine. Again, in the closing chapter, Butler writes (p. 275):-

"We should endeavour to see the so-called inorganic as living in respect of the qualities it has in common with the organic, rather than the organic as non-living in respect of the qualities it has in common with the inorganic."

We conclude our survey of this book by mentioning the literary controversial part chiefly to be found in Chapter IV, but cropping up elsewhere. It refers to interpolations made in the authorised translation of Krause's "Life of Erasmus Darwin." Only one side is presented; and we are not called upon, here or elsewhere, to discuss the merits of the question.

"LUCK, OR CUNNING, as the Main Means of Organic Modification? an Attempt to throw Additional Light upon the late Mr. Charles Darwin's Theory of Natural Selection" (1887), completes the series of biological books. This is mainly a book of strenuous polemic. It brings out still more forcibly the Hering-Butler doctrine of continued personality from generation to generation, and of the working of unconscious memory throughout; and points out that, while this is implicit in much of the teaching of Herbert Spencer, Romanes, and others, it was nowhere--even after the appearance of "Life and Habit"--explicitly recognised by them, but, on the contrary, masked by inconsistent statements and teaching. Not Luck but Cunning, not the uninspired weeding out by Natural Selection but the intelligent striving of the organism, is at the bottom of the useful variety of organic life. And the parallel is drawn that not the happy accident of time and place, but the Machiavellian cunning of Charles Darwin, succeeded in imposing, as entirely his own, on the civilised world an uninspired and inadequate theory of evolution wherein luck played the leading part; while the more inspired and inspiring views of the older evolutionists had failed by the inferiority of their luck. On this controversy I am bound to say that I do not in the very least share Butler's opinions; and I must ascribe them to his lack of personal familiarity with the biologists of the day and their modes of thought and of work. Butler everywhere undervalues the important work of elimination played by Natural Selection in its widest sense.

The "Conclusion" of "Luck, or Cunning?" shows a strong advance in monistic views, and a yet more marked development in the vibration hypothesis of memory given by Hering and only adopted with the greatest reserve in "Unconscious Memory."

"Our conception, then, concerning the nature of any matter depends solely upon its kind and degree of unrest, that is to say, on the characteristics of the vibrations that are going on within it. The exterior object vibrating in a certain way imparts some of its vibrations to our brain; but if the state of the thing itself depends upon its vibrations, it [the thing] must be considered as to all intents and purposes the vibrations themselves--plus, of course, the underlying substance that is vibrating. . . . The same vibrations, therefore, form the substance remembered, introduce an infinitesimal dose of it within the brain, modify the substance remembering, and, in the course of time, create and further modify the mechanism of both the sensory and the motor nerves. Thought and thing are one.

"I commend these two last speculations to the reader's charitable consideration, as feeling that I am here travelling beyond the ground on which I can safely venture. . . . I believe they are both substantially true."

In 1885 he had written an abstract of these ideas in his notebooks (see New Quarterly Review, 1910, p. 116), and as in "Luck, or Cunning?" associated them vaguely with the unitary conceptions introduced into chemistry by Newlands and Mendelejeff. Judging himself as an outsider, the author of "Life and Habit" would certainly have considered the mild expression of faith, "I believe they are both substantially true," equivalent to one of extreme doubt. Thus "the fact of the Archbishop's recognising this as among the number of his beliefs is conclusive evidence, with those who have devoted attention to the laws of thought, that his mind is not yet clear" on the matter of the belief avowed (see "Life and Habit," pp. 24, 25).

To sum up: Butler's fundamental attitude to the vibration hypothesis was all through that taken in "Unconscious Memory"; he played with it as a pretty pet, and fancied it more and more as time went on; but instead of backing it for all he was worth, like the main theses of "Life and Habit," he put a big stake on it--and then hedged.

The last of Butler's biological writings is the Essay, "THE DEADLOCK IN DARWINISM," containing much valuable criticism on Wallace and Weismann. It is in allusion to the misnomer of Wallace's book, "Darwinism," that he introduces the term "Wallaceism" {0d} for a theory of descent that excludes the transmission of acquired characters. This was, indeed, the chief factor that led Charles Darwin to invent his hypothesis of pangenesis, which, unacceptable as it has proved, had far more to recommend it as a formal hypothesis than the equally formal germ-plasm hypothesis of Weismann.

The chief difficulty in accepting the main theses of Butler and Hering is one familiar to every biologist, and not at all difficult to understand by the layman. Everyone knows that the complicated beings that we term "Animals" and "Plants," consist of a number of more or less individualised units, the cells, each analogous to a simpler being, a Protist--save in so far as the character of the cell unit of the Higher being is modified in accordance with the part it plays in that complex being as a whole. Most people, too, are familiar with the fact that the complex being starts as a single cell, separated from its parent; or, where bisexual reproduction occurs, from a cell due to the fusion of two cells, each detached from its parent. Such cells are called "Germ-cells." The germ-cell, whether of single or of dual origin, starts by dividing repeatedly, so as to form the PRIMARY EMBRYONIC CELLS, a complex mass of cells, at first essentially similar, which, however, as they go on multiplying, undergo differentiations and migrations, losing their simplicity as they do so. Those cells that are modified to take part in the proper work of the whole are called tissue-cells. In virtue of their activities, their growth and reproductive power are limited--much more in Animals than in Plants, in Higher than in Lower beings. It is these tissues, or some of them, that receive the impressions from the outside which leave the imprint of memory. Other cells, which may be closely associated into a continuous organ, or more or less surrounded by tissue-cells, whose part it is to nourish them, are called "secondary embryonic cells," or "germ-cells." The germ-cells may be differentiated in the young organism at a very early stage, but in Plants they are separated at a much later date from the less isolated embryonic regions that provide for the Plant's branching; in all cases we find embryonic and germ-cells screened from the life processes of the complex organism, or taking no very obvious part in it, save to form new tissues or new organs, notably in Plants.

Again, in ourselves, and to a greater or less extent in all Animals, we find a system of special tissues set apart for the reception and storage of impressions from the outer world, and for guiding the other organs in their appropriate responses--the "Nervous System"; and when this system is ill-developed or out of gear the remaining organs work badly from lack of proper skilled guidance and co-ordination. How can we, then, speak of "memory" in a germ-cell which has been screened from the experiences of the organism, which is too simple in structure to realise them if it were exposed to them? My own answer is that we cannot form any theory on the subject, the only question is whether we have any right to INFER this "memory" from the BEHAVIOUR of living beings; and Butler, like Hering, Haeckel, and some more modern authors, has shown that the inference is a very strong presumption. Again, it is easy to over-value such complex instruments as we possess. The possessor of an up-to-date camera, well instructed in the function and manipulation of every part, but ignorant of all optics save a hand-to-mouth knowledge of the properties of his own lens, might say that a priori no picture could be taken with a cigar-box perforated by a pin-hole; and our ignorance of the mechanism of the Psychology of any organism is greater by many times than that of my supposed photographer. We know that Plants are able to do many things that can only be accounted for by ascribing to them a "psyche," and these co-ordinated enough to satisfy their needs; and yet they possess no central organ comparable to the brain, no highly specialised system for intercommunication like our nerve trunks and fibres. As Oscar Hertwig says, we are as ignorant of the mechanism of the development of the individual as we are of that of hereditary transmission of acquired characters, and the absence of such mechanism in either case is no reason for rejecting the proven fact.

However, the relations of germ and body just described led Jager, Nussbaum, Galton, Lankester, and, above all, Weismann, to the view that the germ-cells or "stirp" (Galton) were IN the body, but not OF it. Indeed, in the body and out of it, whether as reproductive cells set free, or in the developing embryo, they are regarded as forming one continuous homogeneity, in contrast to the differentiation of the body; and it is to these cells, regarded as a continuum, that the terms stirp, germ-plasm, are especially applied. Yet on this view, so eagerly advocated by its supporters, we have to substitute for the hypothesis of memory, which they declare to have no real meaning here, the far more fantastic hypotheses of Weismann: by these they explain the process of differentiation in the young embryo into new germ and body; and in the young body the differentiation of its cells, each in due time and place, into the varied tissue cells and organs. Such views might perhaps be acceptable if it could be shown that over each cell-division there presided a wise all-guiding genie of transcending intellect, to which Clerk-Maxwell's sorting demons were mere infants. Yet these views have so enchanted many distinguished biologists, that in dealing with the subject they have actually ignored the existence of equally able workers who hesitate to share the extremest of their views. The phenomenon is one well known in hypnotic practice. So long as the non-Weismannians deal with matters outside this discussion, their existence and their work is rated at its just value; but any work of theirs on this point so affects the orthodox Weismannite (whether he accept this label or reject it does not matter), that for the time being their existence and the good work they have done are alike non-existent. {0e}

Butler founded no school, and wished to found none. He desired that
what was true in his work should prevail, and he looked forward
calmly to the time when the recognition of that truth and of his
share in advancing it should give him in the lives of others that
immortality for which alone he craved.

Lamarckian views have never lacked defenders here and in America. Of
the English, Herbert Spencer, who however, was averse to the
vitalistic attitude, Vines and Henslow among botanists, Cunningham
among zoologists, have always resisted Weismannism; but, I think,
none of these was distinctly influenced by Hering and Butler. In
America the majority of the great school of palaeontologists have
been strong Lamarckians, notably Cope, who has pointed out, moreover,
that the transformations of energy in living beings are peculiar to
them.

We have already adverted to Haeckel's acceptance and development of
Hering's ideas in his "Perigenese der Plastidule." Oscar Hertwig has
been a consistent Lamarckian, like Yves Delage of the Sorbonne, and
these occupy pre-eminent positions not only as observers, but as
discriminating theorists and historians of the recent progress of
biology. We may also cite as a Lamarckian--of a sort--Felix Le
Dantec, the leader of the chemico-physical school of the present day.

But we must seek elsewhere for special attention to the points which
Butler regarded as the essentials of "Life and Habit." In 1893 Henry
P. Orr, Professor of Biology in the University of Louisiana,
published a little book entitled "A Theory of Heredity." Herein he
insists on the nervous control of the whole body, and on the
transmission to the reproductive cells of such stimuli, received by
the body, as will guide them on their path until they shall have
acquired adequate experience of their own in the new body they have

formed. I have found the name of neither Butler nor Hering, but the treatment is essentially on their lines, and is both clear and interesting.

In 1896 I wrote an essay on "The Fundamental Principles of Heredity," primarily directed to the man in the street. This, after being held over for more than a year by one leading review, was "declined with regret," and again after some weeks met the same fate from another editor. It appeared in the pages of "Natural Science" for October, 1897, and in the "Biologisches Centralblatt" for the same year. I reproduce its closing paragraph:-

"This theory [Hering-Butler's] has, indeed, a tentative character, and lacks symmetrical completeness, but is the more welcome as not aiming at the impossible. A whole series of phenomena in organic beings are correlated under the term of MEMORY, CONSCIOUS AND UNCONSCIOUS, PATENT AND LATENT. . . . Of the order of unconscious memory, latent till the arrival of the appropriate stimulus, is all the co-operative growth and work of the organism, including its development from the reproductive cells. Concerning the modus operandi we know nothing: the phenomena may be due, as Hering suggests, to molecular vibrations, which must be at least as distinct from ordinary physical disturbances as Rontgen's rays are from ordinary light; or it may be correlated, as we ourselves are inclined to think, with complex chemical changes in an intricate but orderly succession. For the present, at least, the problem of heredity can only be elucidated by the light of mental, and not material processes."

It will be seen that I express doubts as to the validity of Hering's invocation of molecular vibrations as the mechanism of memory, and suggest as an alternative rhythmic chemical changes. This view has recently been put forth in detail by J. J. Cunningham in his essay on the "Hormone {0f} Theory of Heredity," in the Archiv fur Entwicklungsmechanik (1909), but I have failed to note any direct effect of my essay on the trend of biological thought.

Among post-Darwinian controversies the one that has latterly assumed the greatest prominence is that of the relative importance of small variations in the way of more or less "fluctuations," and of "discontinuous variations," or "mutations," as De Vries has called them. Darwin, in the first four editions of the "Origin of Species," attached more importance to the latter than in subsequent editions; he was swayed in his attitude, as is well known, by an article of the physicist, Fleeming Jenkin, which appeared in the North British Review. The mathematics of this article were unimpeachable, but they were founded on the assumption that exceptional variations would only occur in single individuals, which is, indeed, often the case among those domesticated races on which Darwin especially studied the phenomena of variation. Darwin was no mathematician or physicist, and we are told in his biography that he regarded every tool-shop rule or optician's thermometer as an instrument of precision: so he appears to have regarded Fleeming Jenkin's demonstration as a mathematical deduction which he was bound to accept without criticism.

Mr. William Bateson, late Professor of Biology in the University of Cambridge, as early as 1894 laid great stress on the importance of discontinuous variations, collecting and collating the known facts in his "Materials for the Study of Variations"; but this important work, now become rare and valuable, at the time excited so little interest as to be 'remaindered' within a very few years after publication.

In 1901 Hugo De Vries, Professor of Botany in the University of

Amsterdam, published "Die Mutationstheorie," wherein he showed that mutations or discontinuous variations in various directions may appear simultaneously in many individuals, and in various directions. In the gardener's phrase, the species may take to sporting in various directions at the same time, and each sport may be represented by numerous specimens.

De Vries shows the probability that species go on for long periods showing only fluctuations, and then suddenly take to sporting in the way described, short periods of mutation alternating with long intervals of relative constancy. It is to mutations that De Vries and his school, as well as Luther Burbank, the great former of new fruit- and flower-plants, look for those variations which form the material of Natural Selection. In "God the Known and God the Unknown," which appeared in the Examiner (May, June, and July), 1879, but though then revised was only published posthumously in 1909, Butler anticipates this distinction:-

"Under these circumstances organism must act in one or other of these two ways: it must either change slowly and continuously with the surroundings, paying cash for everything, meeting the smallest change with a corresponding modification, so far as is found convenient, or it must put off change as long as possible, and then make larger and more sweeping changes.

"Both these courses are the same in principle, the difference being one of scale, and the one being a miniature of the other, as a ripple is an Atlantic wave in little; both have their advantages and disadvantages, so that most organisms will take the one course for one set of things and the other for another. They will deal promptly with things which they can get at easily, and which lie more upon the surface; THOSE, HOWEVER, WHICH ARE MORE TROUBLESOME TO

REACH, AND LIE
DEEPER, WILL BE HANDLED UPON MORE CATACLYSMIC PRINCIPLES, BEING
ALLOWED LONGER PERIODS OF REPOSE FOLLOWED BY SHORT PERI-
ODS OF GREATER
ACTIVITY . . . it may be questioned whether what is called a sport is
not the organic expression of discontent which has been long felt,
but which has not been attended to, nor been met step by step by as
much small remedial modification as was found practicable: so that
when a change does come it comes by way of revolution. Or, again
(only that it comes to much the same thing), it may be compared to
one of those happy thoughts which sometimes come to us unbidden after
we have been thinking for a long time what to do, or how to arrange
our ideas, and have yet been unable to come to any conclusion" (pp.
14, 15). {0g}

We come to another order of mind in Hans Driesch. At the time he
began his work biologists were largely busy in a region indicated by
Darwin, and roughly mapped out by Haeckel--that of phylogeny. From
the facts of development of the individual, from the comparison of
fossils in successive strata, they set to work at the construction of
pedigrees, and strove to bring into line the principles of
classification with the more or less hypothetical "stemtrees."
Driesch considered this futile, since we never could reconstruct from
such evidence anything certain in the history of the past. He
therefore asserted that a more complete knowledge of the physics and
chemistry of the organic world might give a scientific explanation of
the phenomena, and maintained that the proper work of the biologist
was to deepen our knowledge in these respects. He embodied his
views, seeking the explanation on this track, filling up gaps and
tracing projected roads along lines of probable truth in his
"Analytische Theorie der organische Entwicklung." But his own work
convinced him of the hopelessness of the task he had undertaken, and

he has become as strenuous a vitalist as Butler. The most complete statement of his present views is to be found in "The Philosophy of Life" (1908-9), being the Giffold Lectures for 1907-8. Herein he postulates a quality ("psychoid") in all living beings, directing energy and matter for the purpose of the organism, and to this he applies the Aristotelian designation "Entelechy." The question of the transmission of acquired characters is regarded as doubtful, and he does not emphasise--if he accepts--the doctrine of continuous personality. His early youthful impatience with descent theories and hypotheses has, however, disappeared.

In the next work the influence of Hering and Butler is definitely present and recognised. In 1906 Signor Eugenio Rignano, an engineer keenly interested in all branches of science, and a little later the founder of the international review, Rivista di Scienza (now simply called Scientia), published in French a volume entitled "Sur la transmissibilite des Caracteres acquis--Hypothese d'un Centro-epigenese." Into the details of the author's work we will not enter fully. Suffice it to know that he accepts the Hering-Butler theory, and makes a distinct advance on Hering's rather crude hypothesis of persistent vibrations by suggesting that the remembering centres store slightly different forms of energy, to give out energy of the same kind as they have received, like electrical accumulators. The last chapter, "Le Phenomene mnemonique et le Phenomene vital," is frankly based on Hering.

In "The Lesson of Evolution" (1907, posthumous, and only published for private circulation) Frederick Wollaston Hutton, F.R.S., late Professor of Biology and Geology, first at Dunedin and after at Christchurch, New Zealand, puts forward a strongly vitalistic view, and adopts Hering's teaching. After stating this he adds, "The same idea of heredity being due to unconscious memory was advocated by Mr. Samuel Butler in his "Life and Habit."

Dr. James Mark Baldwin, Stuart Professor of Psychology in Princeton University, U.S.A., called attention early in the 90's to a reaction characteristic of all living beings, which he terms the "Circular Reaction." We take his most recent account of this from his "Development and Evolution" (1902):- {0h}

"The general fact is that the organism reacts by concentration upon the locality stimulated for the CONTINUANCE of the conditions, movements, stimulations, WHICH ARE VITALLY BENEFICIAL, and for the cessation of the conditions, movements, stimulations WHICH ARE VITALLY DEPRESSING."

This amounts to saying in the terminology of Jenning (see below) that the living organism alters its "physiological states" either for its direct benefit, or for its indirect benefit in the reduction of harmful conditions.

Again:-

"This form of concentration of energy on stimulated localities, with the resulting renewal through movement of conditions that are pleasure-giving and beneficial, and the consequent repetition of the movements is called 'circular reaction.'"

Of course, the inhibition of such movements as would be painful on repetition is merely the negative case of the circular reaction. We must not put too much of our own ideas into the author's mind; he nowhere says explicitly that the animal or plant shows its sense and does this because it likes the one thing and wants it repeated, or dislikes the other and stops its repetition, as Butler would have said. Baldwin is very strong in insisting that no full explanation can be given of living processes, any more than of history, on purely chemico-physical grounds.

The same view is put differently and independently by H. S. Jennings, {0i} who started his investigations of living Protista, the simplest of living beings, with the idea that only accurate and ample observation was needed to enable us to explain all their activities on a mechanical basis, and devised ingenious models of protoplastic movements. He was led, like Driesch, to renounce such efforts as illusory, and has come to the conviction that in the behaviour of these lowly beings there is a purposive and a tentative character--a method of "trial and error"--that can only be interpreted by the invocation of psychology. He points out that after stimulation the "state" of the organism may be altered, so that the response to the same stimulus on repetition is other. Or, as he puts it, the first stimulus has caused the organism to pass into a new "physiological state." As the change of state from what we may call the "primary indifferent state" is advantageous to the organism, we may regard this as equivalent to the doctrine of the "circular reaction," and also as containing the essence of Semon's doctrine of "engrams" or imprints which we are about to consider. We cite one passage which for audacity of thought (underlying, it is true, most guarded expression) may well compare with many of the boldest flights in "Life and Habit":-

"It may be noted that regulation in the manner we have set forth is what, in the behaviour of higher organisms, at least, is called intelligence [the examples have been taken from Protista, Corals, and the Lowest Worms]. If the same method of regulation is found in other fields, there is no reason for refusing to compare the action to intelligence. Comparison of the regulatory processes that are shown in internal physiological changes and in regeneration to intelligence seems to be looked upon sometimes as heretical and unscientific. Yet intelligence is a name applied to processes that actually exist in the regulation of movements, and there is, a priori, no reason why similar processes should not occur in regulation in other fields. When we analyse regulation objectively there seems indeed reason to think that the processes are of the same character in behaviour as elsewhere. If the term intelligence be reserved for the subjective accompaniments of such regulation, then of course we have no direct knowledge of its existence in any of the fields of regulation outside of the self, and in the self perhaps only in behaviour. But in a purely objective consideration there seems no reason to suppose that regulation in behaviour (intelligence) is of a fundamentally different character from regulation elsewhere." ("Method of Regulation," p. 492.)

Jennings makes no mention of questions of the theory of heredity. He has made some experiments on the transmission of an acquired character in Protozoa; but it was a mutilation-character, which is, as has been often shown, {0j} not to the point.

One of the most obvious criticisms of Hering's exposition is based upon the extended use he makes of the word "Memory": this he had foreseen and deprecated.

"We have a perfect right," he says, "to extend our conception of memory so as to make it embrace involuntary [and also unconscious] reproductions of sensations, ideas, perceptions, and efforts; but we find, on having done so, that we have so far enlarged her boundaries that she proves to be an ultimate and original power, the source and, at the same time, the unifying bond, of our whole conscious life." ("Unconscious Memory," p. 68.)

This sentence, coupled with Hering's omission to give to the concept of memory so enlarged a new name, clear alike of the limitations and of the stains of habitual use, may well have been the inspiration of the next work on our list. Richard Semon is a professional zoologist and anthropologist of such high status for his original observations and researches in the mere technical sense, that in these countries he would assuredly have been acclaimed as one of the Fellows of the Royal Society who were Samuel Butler's special aversion. The full title of his book is "DIE MNEME als erhaltende Prinzip im Wechsel des organischen Geschehens" (Munich, Ed. 1, 1904; Ed. 2, 1908). We may translate it "MNEME, a Principle of Conservation in the Transformations of Organic Existence."

From this I quote in free translation the opening passage of Chapter II:-

"We have shown that in very many cases, whether in Protist, Plant, or Animal, when an organism has passed into an indifferent state after the reaction to a stimulus has ceased, its irritable substance has suffered a lasting change: I call this after-action of the stimulus its 'imprint' or 'engraphic' action, since it penetrates and imprints itself in the organic substance; and I term the change so effected an 'imprint' or 'engram' of the stimulus; and the sum of all the imprints possessed by the organism may be called its 'store of imprints,' wherein we must distinguish between those which it has inherited from its forbears and those which it has acquired itself.

Any phenomenon displayed by an organism as the result either of a single imprint or of a sum of them, I term a 'mnemic phenomenon'; and the mnemic possibilities of an organism may be termed, collectively, its 'MNEME.'

"I have selected my own terms for the concepts that I have just defined. On many grounds I refrain from making any use of the good German terms 'Gedachtniss, Erinnerungsbild.' The first and chiefest ground is that for my purpose I should have to employ the German words in a much wider sense than what they usually convey, and thus leave the door open to countless misunderstandings and idle controversies. It would, indeed, even amount to an error of fact to give to the wider concept the name already current in the narrower sense--nay, actually limited, like 'Erinnerungsbild,' to phenomena of consciousness. . . . In Animals, during the course of history, one set of organs has, so to speak, specialised itself for the reception and transmission of stimuli--the Nervous System. But from this specialisation we are not justified in ascribing to the nervous system any monopoly of the function, even when it is as highly developed as in Man. . . . Just as the direct excitability of the nervous system has progressed in the history of the race, so has its capacity for receiving imprints; but neither susceptibility nor retentiveness is its monopoly; and, indeed, retentiveness seems inseparable from susceptibility in living matter."

Semen here takes the instance of stimulus and imprint actions affecting the nervous system of a dog

"who has up till now never experienced aught but kindness from the Lord of Creation, and then one day that he is out alone is pelted with stones by a boy. . . . Here he is affected at once by two sets

of stimuli: (1) the optic stimulus of seeing the boy stoop for
stones and throw them, and (2) the skin stimulus of the pain felt
when they hit him. Here both stimuli leave their imprints; and the
organism is permanently changed in relation to the recurrence of the
stimuli. Hitherto the sight of a human figure quickly stooping had
produced no constant special reaction. Now the reaction is constant,
and may remain so till death. . . . The dog tucks in its tail
between its legs and takes flight, often with a howl [as of] pain."

"Here we gain on one side a deeper insight into the imprint action of
stimuli. It reposes on the lasting change in the conditions of the
living matter, so that the repetition of the immediate or synchronous
reaction to its first stimulus (in this case the stooping of the boy,
the flying stones, and the pain on the ribs), no longer demands, as
in the original state of indifference, the full stimulus a, but may
be called forth by a partial or different stimulus, b (in this case
the mere stooping to the ground). I term the influences by which
such changed reaction are rendered possible, 'outcome-reactions,' and
when such influences assume the form of stimuli, 'outcome-stimuli.'

They are termed "outcome" ("ecphoria") stimuli, because the author
regards them and would have us regard them as the outcome,
manifestation, or efference of an imprint of a previous stimulus. We
have noted that the imprint is equivalent to the changed
"physiological state" of Jennings. Again, the capacity for gaining
imprints and revealing them by outcomes favourable to the individual
is the "circular reaction" of Baldwin, but Semon gives no reference
to either author. {0k}

In the preface to his first edition (reprinted in the second) Semon
writes, after discussing the work of Hering and Haeckel:-

"The problem received a more detailed treatment in Samuel Butler's book, 'Life and Habit,' published in 1878. Though he only made acquaintance with Hering's essay after this publication, Butler gave what was in many respects a more detailed view of the coincidences of these different phenomena of organic reproduction than did Hering. With much that is untenable, Butler's writings present many a brilliant idea; yet, on the whole, they are rather a retrogression than an advance upon Hering. Evidently they failed to exercise any marked influence upon the literature of the day."

This judgment needs a little examination. Butler claimed, justly, that his "Life and Habit" was an advance on Hering in its dealing with questions of hybridity, and of longevity puberty and sterility. Since Semon's extended treatment of the phenomena of crosses might almost be regarded as the rewriting of the corresponding section of "Life and Habit" in the "Mneme" terminology, we may infer that this view of the question was one of Butler's "brilliant ideas." That Butler shrank from accepting such a formal explanation of memory as Hering did with his hypothesis should certainly be counted as a distinct "advance upon Hering," for Semon also avoids any attempt at an explanation of "Mneme." I think, however, we may gather the real meaning of Semon's strictures from the following passages:-

"I refrain here from a discussion of the development of this theory of Lamarck's by those Neo-Lamarckians who would ascribe to the individual elementary organism an equipment of complex psychical powers--so to say, anthropomorphic perception and volitions. This treatment is no longer directed by the scientific principle of referring complex phenomena to simpler laws, of deducing even human intellect and will from simpler elements. On the contrary, they follow that most abhorrent method of taking the most complex and unresolved as a datum, and employing it as an explanation. The adoption of such a method, as formerly by Samuel Butler, and recently by Pauly, I regard as a big and dangerous step backward" (ed. 2, pp. 380-1, note).

Thus Butler's alleged retrogressions belong to the same order of thinking that we have seen shared by Driesch, Baldwin, and Jennings, and most explicitly avowed, as we shall see, by Francis Darwin. Semon makes one rather candid admission, "The impossibility of interpreting the phenomena of physiological stimulation by those of direct reaction, and the undeception of those who had put faith in this being possible, have led many on the BACKWARD PATH OF VITALISM." Semon assuredly will never be able to complete his theory of "Mneme" until, guided by the experience of Jennings and Driesch, he forsakes the blind alley of mechanisticism and retraces his steps to reasonable vitalism.

But the most notable publications bearing on our matter are incidental to the Darwin Celebrations of 1908-9. Dr. Francis Darwin, son, collaborator, and biographer of Charles Darwin, was selected to preside over the Meeting of the British Association held in Dublin in 1908, the jubilee of the first publications on Natural Selection by his father and Alfred Russel Wallace. In this address we find the theory of Hering, Butler, Rignano, and Semon taking its proper place as a vera causa of that variation which Natural Selection must find before it can act, and recognised as the basis of a rational theory of the development of the individual and of the race. The organism is essentially purposive: the impossibility of devising any adequate accounts of organic form and function without taking account of the psychical side is most strenuously asserted. And with our regret that past misunderstandings should be so prominent in Butler's works, it was very pleasant to hear Francis Darwin's quotation from Butler's translation of Hering {0l} followed by a personal tribute to Butler himself.

In commemoration of the centenary of the birth of Charles Darwin and of the fiftieth anniversary of the publication of the "Origin of Species," at the suggestion of the Cambridge Philosophical Society, the University Press published during the current year a volume entitled "Darwin and Modern Science," edited by Mr. A. C. Seward, Professor of Botany in the University. Of the twenty-nine essays by men of science of the highest distinction, one is of peculiar interest to the readers of Samuel Butler: "Heredity and Variation in Modern Lights," by Professor W. Bateson, F.R.S., to whose work on "Discontinuous Variations" we have already referred. Here once more Butler receives from an official biologist of the first rank full recognition for his wonderful insight and keen critical power. This is the more noteworthy because Bateson has apparently no faith in the transmission of acquired characters; but such a passage as this would have commended itself to Butler's admiration:-

"All this indicates a definiteness and specific order in heredity, and therefore in variation. This order cannot by the nature of the case be dependent on Natural Selection for its existence, but must be a consequence of the fundamental chemical and physical nature of living things. The study of Variation had from the first shown that an orderliness of this kind was present. The bodies and properties of living things are cosmic, not chaotic. No matter how low in the scale we go, never do we find the slightest hint of a diminution in that all-pervading orderliness, nor can we conceive an organism existing for one moment in any other state."

We have now before us the materials to determine the problem of Butler's relation to biology and to biologists. He was, we have seen, anticipated by Hering; but his attitude was his own, fresh and original. He did not hamper his exposition, like Hering, by a subsidiary hypothesis of vibrations which may or may not be true, which burdens the theory without giving it greater carrying power or persuasiveness, which is based on no objective facts, and which, as Semon has practically demonstrated, is needless for the detailed working out of the theory. Butler failed to impress the biologists of his day, even those on whom, like Romanes, he might have reasonably counted for understanding and for support. But he kept alive Hering's work when it bade fair to sink into the limbo of obsolete hypotheses. To use Oliver Wendell Holmes's phrase, he "depolarised" evolutionary thought. We quote the words of a young biologist, who, when an ardent and dogmatic Weismannist of the most pronounced type, was induced to read "Life and Habit": "The book was to me a transformation and an inspiration." Such learned writings as Semon's or Hering's could never produce such an effect: they do not penetrate to the heart of man; they cannot carry conviction to the intellect already filled full with rival theories, and with the unreasoned faith that to-morrow or next day a new discovery will obliterate all distinction between Man and his makings. The mind must needs be open for the reception of truth, for the rejection of prejudice; and the violence of a Samuel Butler may in the future as in the past be needed to shatter the coat of mail forged by too exclusively professional a training.

MARCUS HARTOG
Cork, April, 1910

AUTHOR'S PREFACE

Not finding the "well-known German scientific journal Kosmos" {0m} entered in the British Museum Catalogue, I have presented the Museum with a copy of the number for February 1879, which contains the article by Dr. Krause of which Mr. Charles Darwin has given a translation, the accuracy of which is guaranteed--so he informs us-- by the translator's "scientific reputation together with his knowledge of German." {0n}

I have marked the copy, so that the reader can see at a glance what passages has been suppressed and where matter has been interpolated.

I have also present a copy of "Erasmus Darwin." I have marked this too, so that the genuine and spurious passages can be easily distinguished.

I understand that both the "Erasmus Darwin" and the number of Kosmos have been sent to the Keeper of Printed Books, with instructions that they shall be at once catalogued and made accessible to readers, and do not doubt that this will have been done before the present volume is published. The reader, therefore, who may be sufficiently interested in the matter to care to see exactly what has been done will now have an opportunity of doing so.

October 25, 1880.

CHAPTER I

Introduction--General ignorance on the subject of evolution at the time the "Origin of Species" was published in 1859.

There are few things which strike us with more surprise, when we review the course taken by opinion in the last century, than the suddenness with which belief in witchcraft and demoniacal possession came to an end. This has been often remarked upon, but I am not acquainted with any record of the fact as it appeared to those under whose eyes the change was taking place, nor have I seen any contemporary explanation of the reasons which led to the apparently sudden overthrow of a belief which had seemed hitherto to be deeply rooted in the minds of almost all men. As a parallel to this, though in respect of the rapid spread of an opinion, and not its decadence, it is probable that those of our descendants who take an interest in ourselves will note the suddenness with which the theory of evolution, from having been generally ridiculed during a period of over a hundred years, came into popularity and almost universal acceptance among educated people.

It is indisputable that this has been the case; nor is it less indisputable that the works of Mr. Darwin and Mr. Wallace have been the main agents in the change that has been brought about in our opinions. The names of Cobden and Bright do not stand more prominently forward in connection with the repeal of the Corn Laws than do those of Mr. Darwin and Mr. Wallace in connection with the general acceptance of the theory of evolution. There is no living philosopher who has anything like Mr. Darwin's popularity with Englishmen generally; and not only this, but his power of fascination extends all over Europe, and indeed in every country in which civilisation has obtained footing: not among the illiterate masses, though these are rapidly following the suit of the educated classes, but among experts and those who are most capable of judging. France, indeed--the country of Buffon and Lamarck--must be counted an exception to the general rule, but in England and Germany there are few men of scientific reputation who do not accept Mr. Darwin as the founder of what is commonly called "Darwinism," and regard him as perhaps the most penetrative and profound philosopher of modern times.

To quote an example from the last few weeks only, {2} I have observed that Professor Huxley has celebrated the twenty-first year since the "Origin of Species" was published by a lecture at the Royal Institution, and am told that he described Mr. Darwin's candour as something actually "terrible" (I give Professor Huxley's own word, as reported by one who heard it); and on opening a small book entitled "Degeneration," by Professor Ray Lankester, published a few days before these lines were written, I find the following passage amid more that is to the same purport:-

"Suddenly one of those great guesses which occasionally appear in the history of science was given to the science of biology by the imaginative insight of that greatest of living naturalists--I would say that greatest of living men--Charles Darwin."--Degeneration, p. 10.

This is very strong language, but it is hardly stronger than that habitually employed by the leading men of science when they speak of

Mr. Darwin. To go farther afield, in February 1879 the Germans devoted an entire number of one of their scientific periodicals {3} to the celebration of Mr. Darwin's seventieth birthday. There is no other Englishman now living who has been able to win such a compliment as this from foreigners, who should be disinterested judges.

Under these circumstances, it must seem the height of presumption to differ from so great an authority, and to join the small band of malcontents who hold that Mr. Darwin's reputation as a philosopher, though it has grown up with the rapidity of Jonah's gourd, will yet not be permanent. I believe, however, that though we must always gladly and gratefully owe it to Mr. Darwin and Mr. Wallace that the public mind has been brought to accept evolution, the admiration now generally felt for the "Origin of Species" will appear as unaccountable to our descendants some fifty or eighty years hence as the enthusiasm of our grandfathers for the poetry of Dr. Erasmus Darwin does to ourselves; and as one who has yielded to none in respect of the fascination Mr. Darwin has exercised over him, I would fain say a few words of explanation which may make the matter clearer to our future historians. I do this the more readily because I can at the same time explain thus better than in any other way the steps which led me to the theory which I afterwards advanced in "Life and Habit."

This last, indeed, is perhaps the main purpose of the earlier chapters of this book. I shall presently give a translation of a lecture by Professor Ewald Hering of Prague, which appeared ten years ago, and which contains so exactly the theory I subsequently advocated myself, that I am half uneasy lest it should be supposed that I knew of Professor Hering's work and made no reference to it. A friend to whom I submitted my translation in MS., asking him how closely he thought it resembled "Life and Habit," wrote back that it

gave my own ideas almost in my own words. As far as the ideas are
concerned this is certainly the case, and considering that Professor
Hering wrote between seven and eight years before I did, I think it
due to him, and to my readers as well as to myself, to explain the
steps which led me to my conclusions, and, while putting Professor
Hering's lecture before them, to show cause for thinking that I
arrived at an almost identical conclusion, as it would appear, by an
almost identical road, yet, nevertheless, quite independently, I must
ask the reader, therefore, to regard these earlier chapters as in
some measure a personal explanation, as well as a contribution to the
history of an important feature in the developments of the last
twenty years. I hope also, by showing the steps by which I was led
to my conclusions, to make the conclusions themselves more acceptable
and easy of comprehension.

Being on my way to New Zealand when the "Origin of Species" appeared,
I did not get it till 1860 or 1861. When I read it, I found "the
theory of natural selection" repeatedly spoken of as though it were a
synonym for "the theory of descent with modification"; this is
especially the case in the recapitulation chapter of the work. I
failed to see how important it was that these two theories--if indeed
"natural selection" can be called a theory--should not be confounded
together, and that a "theory of descent with modification" might be
true, while a "theory of descent with modification through natural
selection" {4} might not stand being looked into.

If any one had asked me to state in brief what Mr. Darwin's theory
was, I am afraid I might have answered "natural selection," or
"descent with modification," whichever came first, as though the one
meant much the same as the other. I observe that most of the leading
writers on the subject are still unable to catch sight of the
distinction here alluded to, and console myself for my want of acumen
by reflecting that, if I was misled, I was misled in good company.

I--and I may add, the public generally--failed also to see what the
unaided reader who was new to the subject would be almost certain to
overlook. I mean, that, according to Mr. Darwin, the variations
whose accumulation resulted in diversity of species and genus were
indefinite, fortuitous, attributable but in small degree to any known
causes, and without a general principle underlying them which would
cause them to appear steadily in a given direction for many
successive generations and in a considerable number of individuals at
the same time. We did not know that the theory of evolution was one
that had been quietly but steadily gaining ground during the last
hundred years. Buffon we knew by name, but he sounded too like
"buffoon" for any good to come from him. We had heard also of
Lamarck, and held him to be a kind of French Lord Monboddo; but we
knew nothing of his doctrine save through the caricatures promulgated
by his opponents, or the misrepresentations of those who had another
kind of interest in disparaging him. Dr. Erasmus Darwin we believed
to be a forgotten minor poet, but ninety-nine out of every hundred of
us had never so much as heard of the "Zoonomia." We were little
likely, therefore, to know that Lamarck drew very largely from
Buffon, and probably also from Dr. Erasmus Darwin, and that this
last-named writer, though essentially original, was founded upon
Buffon, who was greatly more in advance of any predecessor than any
successor has been in advance of him.

We did not know, then, that according to the earlier writers the
variations whose accumulation results in species were not fortuitous
and definite, but were due to a known principle of universal
application--namely, "sense of need"--or apprehend the difference
between a theory of evolution which has a backbone, as it were, in
the tolerably constant or slowly varying needs of large numbers of
individuals for long periods together, and one which has no such
backbone, but according to which the progress of one generation is

always liable to be cancelled and obliterated by that of the next.
We did not know that the new theory in a quiet way professed to tell
us less than the old had done, and declared that it could throw
little if any light upon the matter which the earlier writers had
endeavoured to illuminate as the central point in their system. We
took it for granted that more light must be being thrown instead of
less; and reading in perfect good faith, we rose from our perusal
with the impression that Mr. Darwin was advocating the descent of all
existing forms of life from a single, or from, at any rate, a very
few primordial types; that no one else had done this hitherto, or
that, if they had, they had got the whole subject into a mess, which
mess, whatever it was--for we were never told this--was now being
removed once for all by Mr. Darwin.

The evolution part of the story, that is to say, the fact of
evolution, remained in our minds as by far the most prominent feature
in Mr. Darwin's book; and being grateful for it, we were very ready
to take Mr. Darwin's work at the estimate tacitly claimed for it by
himself, and vehemently insisted upon by reviewers in influential
journals, who took much the same line towards the earlier writers on
evolution as Mr. Darwin himself had taken. But perhaps nothing more
prepossessed us in Mr. Darwin's favour than the air of candour that
was omnipresent throughout his work. The prominence given to the
arguments of opponents completely carried us away; it was this which
threw us off our guard. It never occurred to us that there might be
other and more dangerous opponents who were not brought forward. Mr.
Darwin did not tell us what his grandfather and Lamarck would have
had to say to this or that. Moreover, there was an unobtrusive
parade of hidden learning and of difficulties at last overcome which
was particularly grateful to us. Whatever opinion might be
ultimately come to concerning the value of his theory, there could be
but one about the value of the example he had set to men of science
generally by the perfect frankness and unselfishness of his work.

Friends and foes alike combined to do homage to Mr. Darwin in this respect.

For, brilliant as the reception of the "Origin of Species" was, it met in the first instance with hardly less hostile than friendly criticism. But the attacks were ill-directed; they came from a suspected quarter, and those who led them did not detect more than the general public had done what were the really weak places in Mr. Darwin's armour. They attacked him where he was strongest; and above all, they were, as a general rule, stamped with a disingenuousness which at that time we believed to be peculiar to theological writers and alien to the spirit of science. Seeing, therefore, that the men of science ranged themselves more and more decidedly on Mr. Darwin's side, while his opponents had manifestly--so far as I can remember, all the more prominent among them--a bias to which their hostility was attributable, we left off looking at the arguments against "Darwinism," as we now began to call it, and pigeon-holed the matter to the effect that there was one evolution, and that Mr. Darwin was its prophet.

The blame of our errors and oversights rests primarily with Mr. Darwin himself. The first, and far the most important, edition of the "Origin of Species" came out as a kind of literary Melchisedec, without father and without mother in the works of other people. Here is its opening paragraph:-

"When on board H.M.S. 'Beagle' as naturalist, I was much struck with certain facts in the distribution of the inhabitants of South America, and in the geological relations of the present to the past inhabitants of that continent. These facts seemed to me to throw some light on the origin of species--that mystery of mysteries, as it has been called by one of our greatest philosophers. On my return home, it occurred to me, in 1837, that something might be made out on this question by patiently accumulating and reflecting upon all sorts of facts which could possibly have any bearing on it. After five years' work I allowed myself to speculate on the subject, and drew up some short notes; these I enlarged in 1844 into a sketch of the conclusions which then seemed to me probable: from that period to the present day I have steadily pursued the same object. I hope that I may be excused for entering on these personal details, as I give them to show that I have not been hasty in coming to a decision." {8a}

In the latest edition this passage remains unaltered, except in one unimportant respect. What could more completely throw us off the scent of the earlier writers? If they had written anything worthy of our attention, or indeed if there had been any earlier writers at all, Mr. Darwin would have been the first to tell us about them, and to award them their due meed of recognition. But, no; the whole thing was an original growth in Mr. Darwin's mind, and he had never so much as heard of his grandfather, Dr. Erasmus Darwin.

Dr. Krause, indeed, thought otherwise. In the number of Kosmos for February 1879 he represented Mr. Darwin as in his youth approaching the works of his grandfather with all the devotion which people usually feel for the writings of a renowned poet. {8b} This should perhaps be a delicately ironical way of hinting that Mr. Darwin did not read his grandfather's books closely; but I hardly think that Dr. Krause looked at the matter in this light, for he goes on to say that "almost every single work of the younger Darwin may be paralleled by at least a chapter in the works of his ancestor: the mystery of heredity, adaptation, the protective arrangements of animals and plants, sexual selection, insectivorous plants, and the analysis of the emotions and sociological impulses; nay, even the studies on infants are to be found already discussed in the pages of the elder Darwin." {8c}

Nevertheless, innocent as Mr. Darwin's opening sentence appeared, it contained enough to have put us upon our guard. When he informed us that, on his return from a long voyage, "it occurred to" him that the way to make anything out about his subject was to collect and reflect upon the facts that bore upon it, it should have occurred to us in our turn, that when people betray a return of consciousness upon such matters as this, they are on the confines of that state in which other and not less elementary matters will not "occur to" them. The introduction of the word "patiently" should have been conclusive. I will not analyse more of the sentence, but will repeat the next two lines:- "After five years of work, I allowed myself to speculate upon the subject, and drew up some short notes." We read this, thousands of us, and were blind.

If Dr. Erasmus Darwin's name was not mentioned in the first edition of the "Origin of Species," we should not be surprised at there being no notice taken of Buffon, or at Lamarck's being referred to only twice--on the first occasion to be serenely waved aside, he and all his works; {9a} on the second, {9b} to be commended on a point of detail. The author of the "Vestiges of Creation" was more widely known to English readers, having written more recently and nearer home. He was dealt with summarily, on an early and prominent page, by a misrepresentation, which was silently expunged in later editions of the "Origin of Species." In his later editions (I believe first in his third, when 6000 copies had been already sold), Mr. Darwin did indeed introduce a few pages in which he gave what he designated as a "brief but imperfect sketch" of the progress of opinion on the origin of species prior to the appearance of his own work; but the general impression which a book conveys to, and leaves upon, the public is conveyed by the first edition--the one which is alone, with rare exceptions, reviewed; and in the first edition of the "Origin of Species" Mr. Darwin's great precursors were all either ignored or misrepresented. Moreover, the "brief but imperfect sketch," when it did come, was so very brief, but, in spite of this (for this is what I suppose Mr. Darwin must mean), so very imperfect, that it might as well have been left unwritten for all the help it gave the reader to see the true question at issue between the original propounders of the theory of evolution and Mr. Charles Darwin himself.

That question is this: Whether variation is in the main attributable to a known general principle, or whether it is not?--whether the minute variations whose accumulation results in specific and generic differences are referable to something which will ensure their appearing in a certain definite direction, or in certain definite directions, for long periods together, and in many individuals, or whether they are not?--whether, in a word, these variations are in the main definite or indefinite?

It is observable that the leading men of science seem rarely to understand this even now. I am told that Professor Huxley, in his recent lecture on the coming of age of the "Origin of Species," never so much as alluded to the existence of any such division of opinion as this. He did not even, I am assured, mention "natural selection," but appeared to believe, with Professor Tyndall, {10a} that "evolution" is "Mr. Darwin's theory." In his article on evolution in the latest edition of the "Encyclopaedia Britannica," I find only a veiled perception of the point wherein Mr. Darwin is at variance with his precursors. Professor Huxley evidently knows little of these writers beyond their names; if he had known more, it is impossible he should have written that "Buffon contributed nothing to the general doctrine of evolution," {10b} and that Erasmus Darwin, "though a zealous evolutionist, can hardly be said to have made any real advance on his predecessors." {11} The article is in a high degree unsatisfactory, and betrays at once an amount of ignorance and of perception which leaves an uncomfortable impression.

If this is the state of things that prevails even now, it is not surprising that in 1860 the general public should, with few exceptions, have known of only one evolution, namely, that propounded by Mr. Darwin. As a member of the general public, at that time residing eighteen miles from the nearest human habitation, and three days' journey on horseback from a bookseller's shop, I became one of Mr. Darwin's many enthusiastic admirers, and wrote a philosophical dialogue (the most offensive form, except poetry and books of travel into supposed unknown countries, that even literature can assume) upon the "Origin of Species." This production appeared in the Press, Canterbury, New Zealand, in 1861 or 1862, but I have long lost the only copy I had.

CHAPTER II

How I came to write "Life and Habit," and the circumstances of its completion.

It was impossible, however, for Mr. Darwin's readers to leave the matter as Mr. Darwin had left it. We wanted to know whence came that germ or those germs of life which, if Mr. Darwin was right, were once the world's only inhabitants. They could hardly have come hither from some other world; they could not in their wet, cold, slimy state have travelled through the dry ethereal medium which we call space, and yet remained alive. If they travelled slowly, they would die; if fast, they would catch fire, as meteors do on entering the earth's atmosphere. The idea, again, of their having been created by a quasi-anthropomorphic being out of the matter upon the earth was at variance with the whole spirit of evolution, which indicated that no such being could exist except as himself the result, and not the cause, of evolution. Having got back from ourselves to the monad, we were suddenly to begin again with something which was either unthinkable, or was only ourselves again upon a larger scale--to return to the same point as that from which we had started, only made harder for us to stand upon.

There was only one other conception possible, namely, that the germs had been developed in the course of time from some thing or things that were not what we called living at all; that they had grown up, in fact, out of the material substances and forces of the world in some manner more or less analogous to that in which man had been developed from themselves.

I first asked myself whether life might not, after all, resolve itself into the complexity of arrangement of an inconceivably intricate mechanism. Kittens think our shoe-strings are alive when they see us lacing them, because they see the tag at the end jump about without understanding all the ins and outs of how it comes to do so. "Of course," they argue, "if we cannot understand how a thing comes to move, it must move of itself, for there can be no motion beyond our comprehension but what is spontaneous; if the motion is spontaneous, the thing moving must he alive, for nothing can move of itself or without our understanding why unless it is alive. Everything that is alive and not too large can be tortured, and perhaps eaten; let us therefore spring upon the tag" and they spring upon it. Cats are above this; yet give the cat something which presents a few more of those appearances which she is accustomed to see whenever she sees life, and she will fall as easy a prey to the power which association exercises over all that lives as the kitten itself. Show her a toy-mouse that can run a few yards after being wound up; the form, colour, and action of a mouse being here, there is no good cat which will not conclude that so many of the appearances of mousehood could not be present at the same time without the presence also of the remainder. She will, therefore, spring upon the toy as eagerly as the kitten upon the tag.

Suppose the toy more complex still, so that it might run a few yards, stop, and run on again without an additional winding up; and suppose it so constructed that it could imitate eating and drinking, and could make as though the mouse were cleaning its face with its paws. Should we not at first be taken in ourselves, and assume the presence of the remaining facts of life, though in reality they were not there? Query, therefore, whether a machine so complex as to be prepared with a corresponding manner of action for each one of the successive emergencies of life as it arose, would not take us in for good and all, and look so much as if it were alive that, whether we liked it or not, we should be compelled to think it and call it so; and whether the being alive was not simply the being an exceedingly complicated machine, whose parts were set in motion by the action upon them of exterior circumstances; whether, in fact, man was not a kind of toy-mouse in the shape of a man, only capable of going for seventy or eighty years, instead of half as many seconds, and as much more versatile as he is more durable? Of course I had an uneasy feeling that if I thus made all plants and men into machines, these machines must have what all other machines have if they are machines at all--a designer, and some one to wind them up and work them; but I thought this might wait for the present, and was perfectly ready then, as now, to accept a designer from without, if the facts upon examination rendered such a belief reasonable.

If, then, men were not really alive after all, but were only machines of so complicated a make that it was less trouble to us to cut the

difficulty and say that that kind of mechanism was "being alive," why
should not machines ultimately become as complicated as we are, or at
any rate complicated enough to be called living, and to be indeed as
living as it was in the nature of anything at all to be? If it was
only a case of their becoming more complicated, we were certainly
doing our best to make them so.

I do not suppose I at that time saw that this view comes to much the
same as denying that there are such qualities as life and
consciousness at all, and that this, again, works round to the
assertion of their omnipresence in every molecule of matter, inasmuch
as it destroys the separation between the organic and inorganic, and
maintains that whatever the organic is the inorganic is also. Deny
it in theory as much as we please, we shall still always feel that an
organic body, unless dead, is living and conscious to a greater or
less degree. Therefore, if we once break down the wall of partition
between the organic and inorganic, the inorganic must be living and
conscious also, up to a certain point.

I have been at work on this subject now for nearly twenty years, what
I have published being only a small part of what I have written and
destroyed. I cannot, therefore, remember exactly how I stood in
1863. Nor can I pretend to see far into the matter even now; for
when I think of life, I find it so difficult, that I take refuge in
death or mechanism; and when I think of death or mechanism, I find it
so inconceivable, that it is easier to call it life again. The only
thing of which I am sure is, that the distinction between the organic
and inorganic is arbitrary; that it is more coherent with our other
ideas, and therefore more acceptable, to start with every molecule as
a living thing, and then deduce death as the breaking up of an
association or corporation, than to start with inanimate molecules
and smuggle life into them; and that, therefore, what we call the
inorganic world must be regarded as up to a certain point living, and

instinct, within certain limits, with consciousness, volition, and power of concerted action. It is only of late, however, that I have come to this opinion.

One must start with a hypothesis, no matter how much one distrusts it; so I started with man as a mechanism, this being the strand of the knot that I could then pick at most easily. Having worked upon it a certain time, I drew the inference about machines becoming animate, and in 1862 or 1863 wrote the sketch of the chapter on machines which I afterwards rewrote in "Erewhon." This sketch appeared in the Press, Canterbury, N.Z., June 13, 1863; a copy of it is in the British Museum.

I soon felt that though there was plenty of amusement to be got out of this line, it was one that I should have to leave sooner or later; I therefore left it at once for the view that machines were limbs which we had made, and carried outside our bodies instead of incorporating them with ourselves. A few days or weeks later than June 13, 1863, I published a second letter in the Press putting this view forward. Of this letter I have lost the only copy I had; I have not seen it for years. The first was certainly not good; the second, if I remember rightly, was a good deal worse, though I believed more in the views it put forward than in those of the first letter. I had lost my copy before I wrote "Erewhon," and therefore only gave a couple of pages to it in that book; besides, there was more amusement in the other view. I should perhaps say there was an intermediate extension of the first letter which appeared in the Reasoner, July 1, 1865.

In 1870 and 1871, when I was writing "Erewhon," I thought the best way of looking at machines was to see them as limbs which we had made and carried about with us or left at home at pleasure. I was not, however, satisfied, and should have gone on with the subject at once if I had not been anxious to write "The Fair Haven," a book which is a development of a pamphlet I wrote in New Zealand and published in London in 1865.

As soon as I had finished this, I returned to the old subject, on which I had already been engaged for nearly a dozen years as continuously as other business would allow, and proposed to myself to see not only machines as limbs, but also limbs as machines. I felt immediately that I was upon firmer ground. The use of the word "organ" for a limb told its own story; the word could not have become so current under this meaning unless the idea of a limb as a tool or machine had been agreeable to common sense. What would follow, then, if we regarded our limbs and organs as things that we had ourselves manufactured for our convenience?

The first question that suggested itself was, how did we come to make them without knowing anything about it? And this raised another, namely, how comes anybody to do anything unconsciously? The answer "habit" was not far to seek. But can a person be said to do a thing by force of habit or routine when it is his ancestors, and not he, that has done it hitherto? Not unless he and his ancestors are one and the same person. Perhaps, then, they ARE the same person after all. What is sameness? I remembered Bishop Butler's sermon on "Personal Identity," read it again, and saw very plainly that if a man of eighty may consider himself identical with the baby from whom he has developed, so that he may say, "I am the person who at six months old did this or that," then the baby may just as fairly claim identity with its father and mother, and say to its parents on being born, "I was you only a few months ago." By parity of reasoning each living form now on the earth must be able to claim identity with each generation of its ancestors up to the primordial cell inclusive.

Again, if the octogenarian may claim personal identity with the infant, the infant may certainly do so with the impregnate ovum from which it has developed. If so, the octogenarian will prove to have been a fish once in this his present life. This is as certain as that he was living yesterday, and stands on exactly the same foundation.

I am aware that Professor Huxley maintains otherwise. He writes: "It is not true, for example, . . . that a reptile was ever a fish, but it is true that the reptile embryo" (and what is said here of the reptile holds good also for the human embryo), "at one stage of its development, is an organism, which, if it had an independent existence, must be classified among fishes." {17}

This is like saying, "It is not true that such and such a picture was rejected for the Academy, but it is true that it was submitted to the President and Council of the Royal Academy, with a view to acceptance at their next forthcoming annual exhibition, and that the President and Council regretted they were unable through want of space, &c., &c." --and as much more as the reader chooses. I shall venture, therefore, to stick to it that the octogenarian was once a fish, or if Professor Huxley prefers it, "an organism which must be classified among fishes."

But if a man was a fish once, he may have been a fish a million times over, for aught he knows; for he must admit that his conscious recollection is at fault, and has nothing whatever to do with the matter, which must be decided, not, as it were, upon his own evidence as to what deeds he may or may not recollect having executed, but by the production of his signatures in court, with satisfactory proof that he has delivered each document as his act and deed.

This made things very much simpler. The processes of embryonic development, and instinctive actions, might be now seen as repetitions of the same kind of action by the same individual in successive generations. It was natural, therefore, that they should come in the course of time to be done unconsciously, and a consideration of the most obvious facts of memory removed all further doubt that habit--which is based on memory--was at the bottom of all the phenomena of heredity.

I had got to this point about the spring of 1874, and had begun to write, when I was compelled to go to Canada, and for the next year and a half did hardly any writing. The first passage in "Life and Habit" which I can date with certainty is the one on page 52, which runs as follows:-

"It is one against legion when a man tries to differ from his own past selves. He must yield or die if he wants to differ widely, so as to lack natural instincts, such as hunger or thirst, and not to gratify them. It is more righteous in a man that he should 'eat strange food,' and that his cheek should 'so much as lank not,' than that he should starve if the strange food be at his command. His past selves are living in him at this moment with the accumulated life of centuries. 'Do this, this, this, which we too have done, and found out profit in it,' cry the souls of his forefathers within him. Faint are the far ones, coming and going as the sound of bells wafted on to a high mountain; loud and clear are the near ones, urgent as an alarm of fire."

This was written a few days after my arrival in Canada, June 1874. I was on Montreal mountain for the first time, and was struck with its extreme beauty. It was a magnificent Summer's evening; the noble St. Lawrence flowed almost immediately beneath, and the vast expanse of country beyond it was suffused with a colour which even Italy cannot

surpass. Sitting down for a while, I began making notes for "Life
and Habit," of which I was then continually thinking, and had written
the first few lines of the above, when the bells of Notre Dame in
Montreal began to ring, and their sound was carried to and fro in a
remarkably beautiful manner. I took advantage of the incident to
insert then and there the last lines of the piece just quoted. I
kept the whole passage with hardly any alteration, and am thus able
to date it accurately.

Though so occupied in Canada that writing a book was impossible, I
nevertheless got many notes together for future use. I left Canada
at the end of 1875, and early in 1876 began putting these notes into
more coherent form. I did this in thirty pages of closely written
matter, of which a pressed copy remains in my commonplace-book. I
find two dates among them--the first, "Sunday, Feb. 6, 1876"; and the
second, at the end of the notes, "Feb. 12, 1876."

From these notes I find that by this time I had the theory contained
in "Life and Habit" completely before me, with the four main
principles which it involves, namely, the oneness of personality
between parents and offspring; memory on the part of offspring of
certain actions which it did when in the persons of its forefathers;
the latency of that memory until it is rekindled by a recurrence of
the associated ideas; and the unconsciousness with which habitual
actions come to be performed.

The first half-page of these notes may serve as a sample, and runs
thus:-

"Those habits and functions which we have in common with the lower
animals come mainly within the womb, or are done involuntarily, as
our [growth of] limbs, eyes, &c., and our power of digesting food,

&c. . . .

"We say of the chicken that it knows how to run about as soon as it
is hatched, . . . but had it no knowledge before it was hatched?

"It knew how to make a great many things before it was hatched.

"It grew eyes and feathers and bones.

"Yet we say it knew nothing about all this.

"After it is born it grows more feathers, and makes its bones larger,
and develops a reproductive system.

"Again we say it knows nothing about all this.

"What then does it know?

"Whatever it does not know so well as to be unconscious of knowing
it.

"Knowledge dwells upon the confines of uncertainty.

"When we are very certain, we do not know that we know. When we will
very strongly, we do not know that we will."

I then began my book, but considering myself still a painter by
profession, I gave comparatively little time to writing, and got on
but slowly. I left England for North Italy in the middle of May 1876
and returned early in August. It was perhaps thus that I failed to
hear of the account of Professor Hering's lecture given by Professor
Ray Lankester in Nature, July 13 1876; though, never at that time

seeing Nature, I should probably have missed it under any circumstances. On my return I continued slowly writing. By August 1877 I considered that I had to all intents and purposes completed my book. My first proof bears date October 13, 1877.

At this time I had not been able to find that anything like what I was advancing had been said already. I asked many friends, but not one of them knew of anything more than I did; to them, as to me, it seemed an idea so new as to be almost preposterous; but knowing how things turn up after one has written, of the existence of which one had not known before, I was particularly careful to guard against being supposed to claim originality. I neither claimed it nor wished for it; for if a theory has any truth in it, it is almost sure to occur to several people much about the same time, and a reasonable person will look upon his work with great suspicion unless he can confirm it with the support of others who have gone before him. Still I knew of nothing in the least resembling it, and was so afraid of what I was doing, that though I could see no flaw in the argument, nor any loophole for escape from the conclusion it led to, yet I did not dare to put it forward with the seriousness and sobriety with which I should have treated the subject if I had not been in continual fear of a mine being sprung upon me from some unexpected quarter. I am exceedingly glad now that I knew nothing of Professor Hering's lecture, for it is much better that two people should think a thing out as far as they can independently before they become aware of each other's works but if I had seen it, I should either, as is most likely, not have written at all, or I should have pitched my book in another key.

Among the additions I intended making while the book was in the press, was a chapter on Mr. Darwin's provisional theory of Pangenesis, which I felt convinced must be right if it was Mr. Darwin's, and which I was sure, if I could once understand it, must have an important bearing on "Life and Habit." I had not as yet seen that the principle I was contending for was Darwinian, not Neo-Darwinian. My pages still teemed with allusions to "natural selection," and I sometimes allowed myself to hope that "Life and Habit" was going to be an adjunct to Darwinism which no one would welcome more gladly than Mr. Darwin himself. At this time I had a visit from a friend, who kindly called to answer a question of mine, relative, if I remember rightly, to "Pangenesis." He came, September 26, 1877. One of the first things he said was, that the theory which had pleased him more than anything he had heard of for some time was one referring all life to memory. I said that was exactly what I was doing myself, and inquired where he had met with his theory. He replied that Professor Ray Lankester had written a letter about it in Nature some time ago, but he could not remember exactly when, and had given extracts from a lecture by Professor Ewald Hering, who had originated the theory. I said I should not look at it, as I had completed that part of my work, and was on the point of going to press. I could not recast my work if, as was most likely, I should find something, when I saw what Professor Hering had said, which would make me wish to rewrite my own book; it was too late in the day and I did not feel equal to making any radical alteration; and so the matter ended with very little said upon either side. I wrote, however, afterwards to my friend asking him to tell me the number of Nature which contained the lecture if he could find it, but he was unable to do so, and I was well enough content.

A few days before this I had met another friend, and had explained to him what I was doing. He told me I ought to read Professor Mivart's "Genesis of Species," and that if I did so I should find there were two sides to "natural selection." Thinking, as so many people do--and no wonder--that "natural selection" and evolution were much the same thing, and having found so many attacks upon evolution produce no effect upon me, I declined to read it. I had as yet no idea that a writer could attack Neo-Darwinism without attacking evolution. But my friend kindly sent me a copy; and when I read it, I found myself in the presence of arguments different from those I had met with hitherto, and did not see my way to answering them. I had, however, read only a small part of Professor Mivart's work, and was not fully awake to the position, when the friend referred to in the preceding paragraph called on me.

When I had finished the "Genesis of Species," I felt that something was certainly wanted which should give a definite aim to the variations whose accumulation was to amount ultimately to specific and generic differences, and that without this there could have been no progress in organic development. I got the latest edition of the "Origin of Species" in order to see how Mr. Darwin met Professor Mivart, and found his answers in many respects unsatisfactory. I had lost my original copy of the "Origin of Species," and had not read the book for some years. I now set about reading it again, and came to the chapter on instinct, where I was horrified to find the following passage:-

"But it would be a serious error to suppose that the greater number of instincts have been acquired by habit in one generation and then transmitted by inheritance to the succeeding generations. It can be clearly shown that the most wonderful instincts with which we are acquainted, namely, those of the hive-bee and of many ants, could not possibly have been acquired by habit." {23a}

This showed that, according to Mr. Darwin, I had fallen into serious error, and my faith in him, though somewhat shaken, was far too great to be destroyed by a few days' course of Professor Mivart, the full importance of whose work I had not yet apprehended. I continued to read, and when I had finished the chapter felt sure that I must indeed have been blundering. The concluding words, "I am surprised that no one has hitherto advanced this demonstrative case of neuter insects against the well-known doctrine of inherited habit as advanced by Lamarck," {23b} were positively awful. There was a quiet consciousness of strength about them which was more convincing than any amount of more detailed explanation. This was the first I had heard of any doctrine of inherited habit as having been propounded by Lamarck (the passage stands in the first edition, "the well-known doctrine of Lamarck," p. 242); and now to find that I had been only busying myself with a stale theory of this long-since exploded charlatan--with my book three parts written and already in the press--it was a serious scare.

On reflection, however, I was again met with the overwhelming weight of the evidence in favour of structure and habit being mainly due to memory. I accordingly gathered as much as I could second-hand of what Lamarck had said, reserving a study of his "Philosophie Zoologique" for another occasion, and read as much about ants and bees as I could find in readily accessible works. In a few days I saw my way again; and now, reading the "Origin of Species" more closely, and I may say more sceptically, the antagonism between Mr. Darwin and Lamarck became fully apparent to me, and I saw how incoherent and unworkable in practice the later view was in comparison with the earlier. Then I read Mr. Darwin's answers to miscellaneous objections, and was met, and this time brought up, by the passage beginning "In the earlier editions of this work," {24a} &c., on which I wrote very severely in "Life and Habit"; {24b} for I felt by this time that the difference of opinion between us was radical, and that the matter must be fought out according to the rules of the game. After this I went through the earlier part of my book, and cut out the expressions which I had used inadvertently, and which were inconsistent with a teleological view. This necessitated only verbal alterations; for, though I had not known it, the spirit of the book was throughout teleological.

I now saw that I had got my hands full, and abandoned my intention of touching upon "Pangenesis." I took up the words of Mr. Darwin quoted above, to the effect that it would be a serious error to ascribe the greater number of instincts to transmitted habit. I wrote chapter xi. of "Life and Habit," which is headed "Instincts as Inherited Memory"; I also wrote the four subsequent chapters, "Instincts of Neuter Insects," "Lamarck and Mr. Darwin," "Mr. Mivart and Mr. Darwin," and the concluding chapter, all of them in the month of October and the early part of November 1877, the complete book leaving the binder's hands December 4, 1877, but, according to trade custom, being dated 1878. It will be seen that these five concluding chapters were rapidly written, and this may account in part for the directness with which I said anything I had to say about Mr. Darwin; partly this, and partly I felt I was in for a penny and might as well be in for a pound. I therefore wrote about Mr. Darwin's work exactly as I should about any one else's, bearing in mind the inestimable services he had undoubtedly--and must always be counted to have-- rendered to evolution.

CHAPTER III

How I came to write "Evolution, Old and New"--Mr Darwin's "brief but imperfect" sketch of the opinions of the writers on evolution who had preceded him--The reception which "Evolution, Old and New," met with.

Though my book was out in 1877, it was not till January 1878 that I took an opportunity of looking up Professor Ray Lankester's account of Professor Hering's lecture. I can hardly say how relieved I was to find that it sprung no mine upon me, but that, so far as I could gather, Professor Hering and I had come to pretty much the same conclusion. I had already found the passage in Dr. Erasmus Darwin which I quoted in "Evolution, Old and New," but may perhaps as well repeat it here. It runs -

"Owing to the imperfection of language, the offspring is termed a new animal; but is, in truth, a branch or elongation of the parent, since a part of the embryon animal is or was a part of the parent, and, therefore, in strict language, cannot be said to be entirely new at the time of its production, and, therefore, it may retain some of the habits of the parent system." {26}

When, then, the Athenaeum reviewed "Life and Habit" (January 26, 1878), I took the opportunity to write to that paper, calling attention to Professor Hering's lecture, and also to the passage just quoted from Dr. Erasmus Darwin. The editor kindly inserted my letter in his issue of February 9, 1878. I felt that I had now done all in the way of acknowledgment to Professor Hering which it was, for the time, in my power to do.

I again took up Mr. Darwin's "Origin of Species," this time, I admit, in a spirit of scepticism. I read his "brief but imperfect" sketch of the progress of opinion on the origin of species, and turned to each one of the writers he had mentioned. First, I read all the parts of the "Zoonomia" that were not purely medical, and was astonished to find that, as Dr. Krause has since said in his essay on Erasmus Darwin, "HE WAS THE FIRST WHO PROPOSED AND PERSISTENT-LY

CARRIED OUT A WELL-ROUNDED THEORY WITH REGARD TO THE DE-
VELOPMENT OF
THE LIVING WORLD" {27} (italics in original).

This is undoubtedly the case, and I was surprised at finding
Professor Huxley say concerning this very eminent man that he could
"hardly be said to have made any real advance upon his predecessors."
Still more was I surprised at remembering that, in the first edition
of the "Origin of Species," Dr. Erasmus Darwin had never been so much
as named; while in the "brief but imperfect" sketch he was dismissed
with a line of half-contemptuous patronage, as though the mingled
tribute of admiration and curiosity which attaches to scientific
prophecies, as distinguished from discoveries, was the utmost he was
entitled to. "It is curious," says Mr. Darwin innocently, in the
middle of a note in the smallest possible type, "how largely my
grandfather, Dr. Erasmus Darwin, anticipated the views and erroneous
grounds of opinion of Lamarck in his 'Zoonomia' (vol. i. pp. 500-
510), published in 1794"; this was all he had to say about the
founder of "Darwinism," until I myself unearthed Dr. Erasmus Darwin,
and put his work fairly before the present generation in "Evolution,
Old and New." Six months after I had done this, I had the
satisfaction of seeing that Mr. Darwin had woke up to the propriety
of doing much the same thing, and that he had published an
interesting and charmingly written memoir of his grandfather, of
which more anon.

Not that Dr. Darwin was the first to catch sight of a complete theory
of evolution. Buffon was the first to point out that, in view of the
known modifications which had been effected among our domesticated
animals and cultivated plants, the ass and the horse should be
considered as, in all probability, descended from a common ancestor;
yet, if this is so, he writes--if the point "were once gained that
among animals and vegetables there had been, I do not say several

species, but even a single one, which had been produced in the course
of direct descent from another species; if, for example, it could be
once shown that the ass was but a degeneration from the horse, then
there is no further limit to be set to the power of Nature, and we
should not be wrong in supposing that, with sufficient time, she has
evolved all other organised forms from one primordial type" {28a} (et
l'on n'auroit pas tort de supposer, que d'un seul etre elle a su
tirer avec le temps tous les autres etres organises).

This, I imagine, in spite of Professor Huxley's dictum, is
contributing a good deal to the general doctrine of evolution; for
though Descartes and Leibnitz may have thrown out hints pointing more
or less broadly in the direction of evolution, some of which
Professor Huxley has quoted, he has adduced nothing approaching to
the passage from Buffon given above, either in respect of the
clearness with which the conclusion intended to be arrived at is
pointed out, or the breadth of view with which the whole ground of
animal and vegetable nature is covered. The passage referred to is
only one of many to the same effect, and must be connected with one
quoted in "Evolution, Old and New," {28b} from p. 13 of Buffon's
first volume, which appeared in 1749, and than which nothing can well
point more plainly in the direction of evolution. It is not easy,
therefore, to understand why Professor Huxley should give 1753-78 as
the date of Buffon's work, nor yet why he should say that Buffon was
"at first a partisan of the absolute immutability of species," {29a}
unless, indeed, we suppose he has been content to follow that very
unsatisfactory writer, Isidore Geoffroy St. Hilaire (who falls into
this error, and says that Buffon's first volume on animals appeared
1753), without verifying him, and without making any reference to
him.

Professor Huxley quotes a passage from the "Palingenesie
Philosophique" of Bonnet, of which he says that, making allowance for

his peculiar views on the subject of generation, they bear no small resemblance to what is understood by "evolution" at the present day. The most important parts of the passage quoted are as follows:-

"Should I be going too far if I were to conjecture that the plants and animals of the present day have arisen by a sort of natural evolution from the organised beings which peopled the world in its original state as it left the hands of the Creator? . . . In the outset organised beings were probably very different from what they are now--as different as the original world is from our present one. We have no means of estimating the amount of these differences, but it is possible that even our ablest naturalist, if transplanted to the original world, would entirely fail to recognise our plants and animals therein." {29b}

But this is feeble in comparison with Buffon, and did not appear till 1769, when Buffon had been writing on evolution for fully twenty years with the eyes of scientific Europe upon him. Whatever concession to the opinion of Buffon Bonnet may have been inclined to make in 1769, in 1764, when he published his "Contemplation de la Nature," and in 1762 when his "Considerations sur les Corps Organes" appeared, he cannot be considered to have been a supporter of evolution. I went through these works in 1878 when I was writing "Evolution, Old and New," to see whether I could claim him as on my side; but though frequently delighted with his work, I found it impossible to press him into my service.

The pre-eminent claim of Buffon to be considered as the father of the modern doctrine of evolution cannot be reasonably disputed, though he was doubtless led to his conclusions by the works of Descartes and Leibnitz, of both of whom he was an avowed and very warm admirer.

His claim does not rest upon a passage here or there, but upon the
spirit of forty quartos written over a period of about as many years.
Nevertheless he wrote, as I have shown in "Evolution, Old and New,"
of set purpose enigmatically, whereas there was no beating about the
bush with Dr. Darwin. He speaks straight out, and Dr. Krause is
justified in saying of him "THAT HE WAS THE FIRST WHO PROPOSED
AND

PERSISTENTLY CARRIED OUT A WELL-ROUNDED THEORY" of evolution.

I now turned to Lamarck. I read the first volume of the "Philosophie
Zoologique," analysed it and translated the most important parts.
The second volume was beside my purpose, dealing as it does rather
with the origin of life than of species, and travelling too fast and
too far for me to be able to keep up with him. Again I was
astonished at the little mention Mr. Darwin had made of this
illustrious writer, at the manner in which he had motioned him away,
as it were, with his hand in the first edition of the "Origin of
Species," and at the brevity and imperfection of the remarks made
upon him in the subsequent historical sketch.

I got Isidore Geoffroy's "Histoire Naturelle Generale," which Mr.
Darwin commends in the note on the second page of the historical
sketch, as giving "an excellent history of opinion" upon the subject
of evolution, and a full account of Buffon's conclusions upon the
same subject. This at least is what I supposed Mr. Darwin to mean.
What he said was that Isidore Geoffroy gives an excellent history of
opinion on the subject of the date of the first publication of
Lamarck, and that in his work there is a full account of Buffon's
fluctuating conclusions upon THE SAME SUBJECT. {31} But Mr. Darwin
is a more than commonly puzzling writer. I read what M. Geoffroy had
to say upon Buffon, and was surprised to find that, after all,
according to M. Geoffroy, Buffon, and not Lamarck, was the founder of
the theory of evolution. His name, as I have already said, was never

mentioned in the first edition of the "Origin of Species."

M. Geoffroy goes into the accusations of having fluctuated in his opinions, which he tells us have been brought against Buffon, and comes to the conclusion that they are unjust, as any one else will do who turns to Buffon himself. Mr. Darwin, however, in the "brief but imperfect sketch," catches at the accusation, and repeats it while saying nothing whatever about the defence. The following is still all he says: "The first author who in modern times has treated" evolution "in a scientific spirit was Buffon. But as his opinions fluctuated greatly at different periods, and as he does not enter on the causes or means of the transformation of species, I need not here enter on details." On the next page, in the note last quoted, Mr. Darwin originally repeated the accusation of Buffon's having been fluctuating in his opinions, and appeared to give it the imprimatur of Isidore Geoffroy's approval; the fact being that Isidore Geoffroy only quoted the accusation in order to refute it; and though, I suppose, meaning well, did not make half the case he might have done, and abounds with misstatements. My readers will find this matter particularly dealt with in "Evolution, Old and New," Chapter X.

I gather that some one must have complained to Mr. Darwin of his saying that Isidore Geoffroy gave an account of Buffon's "fluctuating conclusions" concerning evolution, when he was doing all he knew to maintain that Buffon's conclusions did not fluctuate; for I see that in the edition of 1876 the word "fluctuating" has dropped out of the note in question, and we now learn that Isidore Geoffroy gives "a full account of Buffon's conclusions," without the "fluctuating." But Buffon has not taken much by this, for his opinions are still left fluctuating greatly at different periods on the preceding page, and though he still was the first to treat evolution in a scientific spirit, he still does not enter upon the causes or means of the transformation of species. No one can understand Mr. Darwin who does not collate the different editions of the "Origin of Species" with some attention. When he has done this, he will know what Newton meant by saying he felt like a child playing with pebbles upon the seashore.

One word more upon this note before I leave it. Mr. Darwin speaks of Isidore Geoffroy's history of opinion as "excellent," and his account of Buffon's opinions as "full." I wonder how well qualified he is to be a judge of these matters? If he knows much about the earlier writers, he is the more inexcusable for having said so little about them. If little, what is his opinion worth?

To return to the "brief but imperfect sketch." I do not think I can ever again be surprised at anything Mr. Darwin may say or do, but if I could, I should wonder how a writer who did not "enter upon the causes or means of the transformation of species," and whose opinions "fluctuated greatly at different periods," can be held to have treated evolution "in a scientific spirit." Nevertheless, when I reflect upon the scientific reputation Mr. Darwin has attained, and the means by which he has won it, I suppose the scientific spirit must be much what he here implies. I see Mr. Darwin says of his own father, Dr. Robert Darwin of Shrewsbury, that he does not consider him to have had a scientific mind. Mr. Darwin cannot tell why he does not think his father's mind to have been fitted for advancing science, "for he was fond of theorising, and was incomparably the best observer" Mr. Darwin ever knew. {33a} From the hint given in the "brief but imperfect sketch," I fancy I can help Mr. Darwin to see why he does not think his father's mind to have been a scientific one. It is possible that Dr. Robert Darwin's opinions did not fluctuate sufficiently at different periods, and that Mr. Darwin considered him as having in some way entered upon the causes or means of the transformation of species. Certainly those who read Mr. Darwin's own works attentively will find no lack of fluctuation in his case; and reflection will show them that a theory of evolution which relies mainly on the accumulation of accidental variations comes very close to not entering upon the causes or means of the transformation of species. {33b}

I have shown, however, in "Evolution, Old and New," that the assertion that Buffon does not enter on the causes or means of the transformation of species is absolutely without foundation, and that, on the contrary, he is continually dealing with this very matter, and devotes to it one of his longest and most important chapters, {33c} but I admit that he is less satisfactory on this head than either Dr. Erasmus Darwin or Lamarck.

As a matter of fact, Buffon is much more of a Neo-Darwinian than either Dr. Erasmus Darwin or Lamarck, for with him the variations are sometimes fortuitous. In the case of the dog, he speaks of them as making their appearance "BY SOME CHANCE common enough with Nature," {33d} and being perpetuated by man's selection. This is exactly the "if any slight favourable variation HAPPEN to arise" of Mr. Charles Darwin. Buffon also speaks of the variations among pigeons arising "par hasard." But these expressions are only ships; his main cause of variation is the direct action of changed conditions of existence, while with Dr. Erasmus Darwin and Lamarck the action of the conditions of existence is indirect, the direct action being that of the animals or plants themselves, in consequence of changed sense of need under changed conditions.

I should say that the sketch so often referred to is at first sight now no longer imperfect in Mr. Darwin's opinion. It was "brief but imperfect" in 1861 and in 1866, but in 1876 I see that it is brief only. Of course, discovering that it was no longer imperfect, I expected to find it briefer. What, then, was my surprise at finding that it had become rather longer? I have found no perfectly satisfactory explanation of this inconsistency, but, on the whole, incline to think that the "greatest of living men" felt himself unequal to prolonging his struggle with the word "but," and resolved to lay that conjunction at all hazards, even though the doing so might cost him the balance of his adjectives; for I think he must know that his sketch is still imperfect.

From Isidore Geoffroy I turned to Buffon himself, and had not long to wait before I felt that I was now brought into communication with the master-mind of all those who have up to the present time busied themselves with evolution. For a brief and imperfect sketch of him, I must refer my readers to "Evolution, Old and New."

I have no great respect for the author of the "Vestiges of Creation," who behaved hardly better to the writers upon whom his own work was founded than Mr. Darwin himself has done. Nevertheless, I could not forget the gravity of the misrepresentation with which he was assailed on page 3 of the first edition of the "Origin of Species," nor impugn the justice of his rejoinder in the following year, {34} when he replied that it was to be regretted Mr. Darwin had read his work "almost as much amiss as if, like its declared opponents, he had an interest in misrepresenting it." {35a} I could not, again, forget that, though Mr. Darwin did not venture to stand by the passage in question, it was expunged without a word of apology or explanation of how it was that he had come to write it. A writer with any claim to our consideration will never fall into serious error about another writer without hastening to make a public apology as soon as he becomes aware of what he has done.

Reflecting upon the substance of what I have written in the last few pages, I thought it right that people should have a chance of knowing more about the earlier writers on evolution than they were likely to hear from any of our leading scientists (no matter how many lectures they may give on the coming of age of the "Origin of Species") except Professor Mivart. A book pointing the difference between teleological and non-teleological views of evolution seemed likely to be useful, and would afford me the opportunity I wanted for giving a resume of the views of each one of the three chief founders of the theory, and of contrasting them with those of Mr. Charles Darwin, as well as for calling attention to Professor Hering's lecture. I accordingly wrote "Evolution, Old and New," which was prominently announced in the leading literary periodicals at the end of February, or on the very first days of March 1879, {35b} as "a comparison of the theories of Buffon, Dr. Erasmus Darwin, and Lamarck, with that of Mr. Charles Darwin, with copious extracts from the works of the three first-named writers." In this book I was hardly able to conceal the fact that, in spite of the obligations under which we must always remain to Mr. Darwin, I had lost my respect for him and for his work.

I should point out that this announcement, coupled with what I had written in "Life and Habit," would enable Mr. Darwin and his friends to form a pretty shrewd guess as to what I was likely to say, and to quote from Dr. Erasmus Darwin in my forthcoming book. The announcement, indeed, would tell almost as much as the book itself to those who knew the works of Erasmus Darwin.

As may be supposed, "Evolution, Old and New," met with a very unfavourable reception at the hands of many of its reviewers. The Saturday Review was furious. "When a writer," it exclaimed, "who has not given as many weeks to the subject as Mr. Darwin has given years, is not content to air his own crude though clever fallacies, but assumes to criticise Mr. Darwin with the superciliousness of a young schoolmaster looking over a boy's theme, it is difficult not to take him more seriously than he deserves or perhaps desires. One would think that Mr. Butler was the travelled and laborious observer of Nature, and Mr. Darwin the pert speculator who takes all his facts at secondhand." {36}

The lady or gentleman who writes in such a strain as this should not
be too hard upon others whom she or he may consider to write like
schoolmasters. It is true I have travelled--not much, but still as
much as many others, and have endeavoured to keep my eyes open to the
facts before me; but I cannot think that I made any reference to my
travels in "Evolution, Old and New." I did not quite see what that
had to do with the matter. A man may get to know a good deal without
ever going beyond the four-mile radius from Charing Cross. Much less
did I imply that Mr. Darwin was pert: pert is one of the last words
that can be applied to Mr. Darwin. Nor, again, had I blamed him for
taking his facts at secondhand; no one is to be blamed for this,
provided he takes well-established facts and acknowledges his
sources. Mr. Darwin has generally gone to good sources. The ground
of complaint against him is that he muddied the water after he had
drawn it, and tacitly claimed to be the rightful owner of the spring,
on the score of the damage he had effected.

Notwithstanding, however, the generally hostile, or more or less
contemptuous, reception which "Evolution, Old and New," met with,
there were some reviews--as, for example, those in the Field, {37a}
the Daily Chronicle, {37b} the Athenaeum, {37c} the Journal of
Science, {37d} the British Journal of Homaeopathy, {37e} the Daily
News, {37f} the Popular Science Review {37g}--which were all I could
expect or wish.

CHAPTER IV

The manner in which Mr. Darwin met "Evolution, Old and New."

By far the most important notice of "Evolution, Old and New," was that taken by Mr. Darwin himself; for I can hardly be mistaken in believing that Dr. Krause's article would have been allowed to repose unaltered in the pages of the well-known German scientific journal, Kosmos, unless something had happened to make Mr. Darwin feel that his reticence concerning his grandfather must now be ended

Mr. Darwin, indeed, gives me the impression of wishing me to understand that this is not the case. At the beginning of this year he wrote to me, in a letter which I will presently give in full, that he had obtained Dr. Krause's consent for a translation, and had arranged with Mr. Dallas, before my book was "announced." "I remember this," he continues, "because Mr. Dallas wrote to tell me of the advertisement." But Mr. Darwin is not a clear writer, and it is impossible to say whether he is referring to the announcement of "Evolution, Old and New"--in which case he means that the arrangements for the translation of Dr. Krause's article were made before the end of February 1879, and before any public intimation could have reached him as to the substance of the book on which I was then engaged--or to the advertisements of its being now published, which appeared at the beginning of May; in which case, as I have said above, Mr. Darwin and his friends had for some time had full opportunity of knowing what I was about. I believe, however, Mr. Darwin to intend that he remembered the arrangements having been made before the beginning of May--his use of the word "announced," instead of "advertised," being an accident; but let this pass.

Some time after Mr. Darwin's work appeared in November 1879, I got it, and looking at the last page of the book, I read as follows:-

"They" (the elder Darwin and Lamarck) "explain the adaptation to purpose of organisms by an obscure impulse or sense of what is purpose-like; yet even with regard to man we are in the habit of saying, that one can never know what so-and-so is good for. The purpose-like is that which approves itself, and not always that which is struggled for by obscure impulses and desires. Just in the same way the beautiful is what pleases."

I had a sort of feeling as though the writer of the above might have had "Evolution, Old and New," in his mind, but went on to the next sentence, which ran -

"Erasmus Darwin's system was in itself a most significant first step in the path of knowledge which his grandson has opened up for us, but to wish to revive it at the present day, as has actually been seriously attempted, shows a weakness of thought and a mental anachronism which no one can envy."

"That's me," said I to myself promptly. I noticed also the position in which the sentence stood, which made it both one of the first that would be likely to catch a reader's eye, and the last he would carry away with him. I therefore expected to find an open reply to some parts of "Evolution, Old and New," and turned to Mr. Darwin's preface.

To my surprise, I there found that what I had been reading could not by any possibility refer to me, for the preface ran as follows:-

"In the February number of a well-known German scientific journal, Kosmos, {39} Dr. Ernest Krause published a sketch of the 'Life of Erasmus Darwin,' the author of the 'Zoonomia,' 'Botanic Garden,' and other works. This article bears the title of a 'Contribution to the History of the Descent Theory'; and Dr. Krause has kindly allowed my brother Erasmus and myself to have a translation made of it for publication in this country."

Then came a note as follows:-

"Mr. Dallas has undertaken the translation, and his scientific reputation, together with his knowledge of German, is a guarantee for its accuracy."

I ought to have suspected inaccuracy where I found so much consciousness of accuracy, but I did not. However this may be, Mr. Darwin pins himself down with every circumstance of preciseness to giving Dr. Krause's article as it appeared in Kosmos,--the whole article, and nothing but the article. No one could know this better than Mr. Darwin.

On the second page of Mr. Darwin's preface there is a small-type note saying that my work, "Evolution, Old and New," had appeared since the publication of Dr. Krause's article. Mr. Darwin thus distinctly precludes his readers from supposing that any passage they might meet with could have been written in reference to, or by the light of, my book. If anything appeared condemnatory of that book, it was an undesigned coincidence, and would show how little worthy of consideration I must be when my opinions were refuted in advance by one who could have no bias in regard to them.

Knowing that if the article I was about to read appeared in February, it must have been published before my book, which was not out till three months later, I saw nothing in Mr. Darwin's preface to complain of, and felt that this was only another instance of my absurd vanity having led me to rush to conclusions without sufficient grounds,--as if it was likely, indeed, that Mr. Darwin should think what I had said of sufficient importance to be affected by it. It was plain that some one besides myself, of whom I as yet knew nothing, had been writing about the elder Darwin, and had taken much the same line concerning him that I had done. It was for the benefit of this person, then, that Dr. Krause's paragraph was intended. I returned to a becoming sense of my own insignificance, and began to read what I supposed to be an accurate translation of Dr. Krause's article as it originally appeared, before "Evolution, Old and New," was published.

On pp. 3 and 4 of Dr. Krause's part of Mr. Darwin's book (pp. 133 and 134 of the book itself), I detected a sub-apologetic tone which a little surprised me, and a notice of the fact that Coleridge when writing on Stillingfleet had used the word "Darwinising." Mr. R. Garnett had called my attention to this, and I had mentioned it in "Evolution, Old and New," but the paragraph only struck me as being a little odd.

When I got a few pages farther on (p. 147 of Mr. Darwin's book), I found a long quotation from Buffon about rudimentary organs, which I had quoted in "Evolution, Old and New." I observed that Dr. Krause used the same edition of Buffon that I did, and began his quotation

two lines from the beginning of Buffon's paragraph, exactly as I had
done; also that he had taken his nominative from the omitted part of
the sentence across a full stop, as I had myself taken it. A little
lower I found a line of Buffon's omitted which I had given, but I
found that at that place I had inadvertently left two pair of
inverted commas which ought to have come out, {41} having intended to
end my quotation, but changed my mind and continued it without
erasing the commas. It seemed to me that these commas had bothered
Dr. Krause, and made him think it safer to leave something out, for
the line he omits is a very good one. I noticed that he translated
"Mais comme nous voulons toujours tout rapporter a un certain but,"
"But we, always wishing to refer," &c., while I had it, "But we, ever
on the look-out to refer," &c.; and "Nous ne faisons pas attention
que nous alterons la philosophie," "We fail to see that thus we
deprive philosophy of her true character," whereas I had "We fail to
see that we thus rob philosophy of her true character." This last
was too much; and though it might turn out that Dr. Krause had quoted
this passage before I had done so, had used the same edition as I
had, had begun two lines from the beginning of a paragraph as I had
done, and that the later resemblances were merely due to Mr. Dallas
having compared Dr. Krause's German translation of Buffon with my
English, and very properly made use of it when he thought fit, it
looked prima facie more as though my quotation had been copied in
English as it stood, and then altered, but not quite altered enough.
This, in the face of the preface, was incredible; but so many points
had such an unpleasant aspect, that I thought it better to send for
Kosmos and see what I could make out.

At this time I knew not one word of German. On the same day,
therefore, that I sent for Kosmos I began acquire that language, and
in the fortnight before Kosmos came had got far enough forward for
all practical purposes--that is to say, with the help of a
translation and a dictionary, I could see whether or no a German

passage was the same as what purported to be its translation.

When Kosmos came I turned to the end of the article to see how the sentence about mental anachronism and weakness of thought looked in German. I found nothing of the kind, the original article ended with some innocent rhyming doggerel about somebody going on and exploring something with eagle eye; but ten lines from the end I found a sentence which corresponded with one six pages from the end of the English translation. After this there could be little doubt that the whole of these last six English pages were spurious matter. What little doubt remained was afterwards removed by my finding that they had no place in any part of the genuine article. I looked for the passage about Coleridge's using the word "Darwinising"; it was not to be found in the German. I looked for the piece I had quoted from Buffon about rudimentary organs; but there was nothing of it, nor indeed any reference to Buffon. It was plain, therefore, that the article which Mr. Darwin had given was not the one he professed to be giving. I read Mr. Darwin's preface over again to see whether he left himself any loophole. There was not a chink or cranny through which escape was possible. The only inference that could be drawn was either that some one had imposed upon Mr. Darwin, or that Mr. Darwin, although it was not possible to suppose him ignorant of the interpolations that had been made, nor of the obvious purpose of the concluding sentence, had nevertheless palmed off an article which had been added to and made to attack "Evolution, Old and New," as though it were the original article which appeared before that book was written. I could not and would not believe that Mr. Darwin had condescended to this. Nevertheless, I saw it was necessary to sift the whole matter, and began to compare the German and the English articles paragraph by paragraph.

On the first page I found a passage omitted from the English, which with great labour I managed to get through, and can now translate as follows:-

"Alexander Von Humboldt used to take pleasure in recounting how powerfully Forster's pictures of the South Sea Islands and St. Pierre's illustrations of Nature had provoked his ardour for travel and influenced his career as a scientific investigator. How much more impressively must the works of Dr. Erasmus Darwin, with their reiterated foreshadowing of a more lofty interpretation of Nature, have affected his grandson, who in his youth assuredly approached them with the devotion due to the works of a renowned poet." {43}

I then came upon a passage common to both German and English, which in its turn was followed in the English by the sub-apologetic paragraph which I had been struck with on first reading, and which was not in the German, its place being taken by a much longer passage which had no place in the English. A little farther on I was amused at coming upon the following, and at finding it wholly transformed in the supposed accurate translation

"How must this early and penetrating explanation of rudimentary organs have affected the grandson when he read the poem of his ancestor! But indeed the biological remarks of this accurate observer in regard to certain definite natural objects must have produced a still deeper impression upon him, pointing, as they do, to questions which hay attained so great a prominence at the present day; such as, Why is any creature anywhere such as we actually see it and nothing else? Why has such and such a plant poisonous juices? Why has such and such another thorns? Why have birds and fishes light-coloured breasts and dark backs, and, Why does every creature resemble the one from which it sprung?" {44a}

I will not weary the reader with further details as to the omissions
from and additions to the German text. Let it suffice that the so-
called translation begins on p. 131 and ends on p. 216 of Mr.
Darwin's book. There is new matter on each one of the pp. 132-139,
while almost the whole of pp. 147-152 inclusive, and the whole of pp.
211-216 inclusive, are spurious--that is to say, not what the purport
to be, not translations from an article that was published in
February 1879, and before "Evolution, Old and New," but
interpolations not published till six months after that book.

Bearing in mind the contents of two of the added passages and the
tenor of the concluding sentence quoted above, {44b} I could no
longer doubt that the article had been altered by the light of and
with a view to "Evolution, Old and New."

The steps are perfectly clear. First Dr. Krause published his
article in Kosmos and my book was announced (its purport being thus
made obvious), both in the month of February 1879. Soon afterwards
arrangements were made for a translation of Dr. Krause's essay, and
were completed by the end of April. Then my book came out, and in
some way or other Dr. Krause happened to get hold of it. He helped
himself--not to much, but to enough; made what other additions to and
omissions from his article he thought would best meet "Evolution, Old
and New," and then fell to condemning that book in a finale that was
meant to be crushing. Nothing was said about the revision which Dr.
Krause's work had undergone, but it was expressly and particularly
declared in the preface that the English translation was an accurate
version of what appeared in the February number of Kosmos, and no
less expressly and particularly stated that my book was published
subsequently to this. Both these statements are untrue; they are in
Mr. Darwin's favour and prejudicial to myself.

All this was done with that well-known "happy simplicity" of which the Pall Mall Gazette, December 12, 1879, declared that Mr. Darwin was "a master." The final sentence, about the "weakness of thought and mental anachronism which no one can envy," was especially successful. The reviewer in the Pall Mall Gazette just quoted from gave it in full, and said that it was thoroughly justified. He then mused forth a general gnome that the "confidence of writers who deal in semi-scientific paradoxes is commonly in inverse proportion to their grasp of the subject." Again my vanity suggested to me that I was the person for whose benefit this gnome was intended. My vanity, indeed, was well fed by the whole transaction; for I saw that not only did Mr. Darwin, who should be the best judge, think my work worth notice, but that he did not venture to meet it openly. As for Dr. Krause's concluding sentence, I thought that when a sentence had been antedated the less it contained about anachronism the better.

Only one of the reviews that I saw of Mr. Darwin's "Life of Erasmus Darwin" showed any knowledge of the facts. The Popular Science Review for January 1880, in flat contradiction to Mr. Darwin's preface, said that only part of Dr. Krause's article was being given by Mr. Darwin. This reviewer had plainly seen both Kosmos and Mr. Darwin's book.

In the same number of the Popular Science Review, and immediately following the review of Mr. Darwin's book, there is a review of "Evolution, Old and New." The writer of this review quotes the passage about mental anachronism as quoted by the reviewer in the Pall Mall Gazette, and adds immediately: "This anachronism has been committed by Mr. Samuel Butler in a . . . little volume now before us, and it is doubtless to this, WHICH APPEARED WHILE HIS OWN WORK WAS IN PROGRESS [italics mine] that Dr. Krause alludes in the foregoing passage." Considering that the editor of the Popular

Science Review and the translator of Dr. Krause's article for Mr. Darwin are one and the same person, it is likely the Popular Science Review is well informed in saying that my book appeared before Dr. Krause's article had been transformed into its present shape, and that my book was intended by the passage in question.

Unable to see any way of escaping from a conclusion which I could not willingly adopt, I thought it best to write to Mr. Darwin, stating the facts as they appeared to myself, and asking an explanation, which I would have gladly strained a good many points to have accepted. It is better, perhaps, that I should give my letter and Darwin's answer in full. My letter ran thus:-

January 2, 1880.

CHARLES DARWIN, ESQ., F.R.S., &c.

Dear Sir,--Will you kindly refer me to the edition of Kosmos which contains the text of Dr. Krause's article on Dr. Erasmus Darwin, as translated by Mr. W. S. Dallas?

I have before me the last February number of Kosmos, which appears by your preface to be the one from which Mr. Dallas has translated, but his translation contains long and important passages which are not in the February number of Kosmos, while many passages in the original article are omitted in the translation.

Among the passages introduced are the last six pages of the English article, which seem to condemn by anticipation the position I have taken as regards Dr. Erasmus Darwin in my book, "Evolution, Old and New," and which I believe I was the first to take. The concluding, and therefore, perhaps, most prominent sentence of the translation you have given to the public stands thus:-

"Erasmus Darwin's system was in itself a most significant first step in the path of knowledge which his grandson has opened up for us, but to wish to revive it at the present day, as has actually been seriously attempted, shows a weakness of thought and a mental anachronism which no man can envy."

The Kosmos which has been sent me from Germany contains no such passage.

As you have stated in your preface that my book, "Evolution, Old and New," appeared subsequently to Dr. Krause's article, and as no intimation is given that the article has been altered and added to since its original appearance, while the accuracy of the translation as though from the February number of Kosmos is, as you expressly say, guaranteed by Mr. Dallas's "scientific reputation together with his knowledge of German," your readers will naturally suppose that all they read in the translation appeared in February last, and therefore before "Evolution, Old and New," was written, and therefore independently of, and necessarily without reference to, that book.

I do not doubt that this was actually the case, but have failed to obtain the edition which contains the passage above referred to, and several others which appear in the translation.

I have a personal interest in this matter, and venture, therefore, to ask for the explanation which I do not doubt you will readily give me.--Yours faithfully,

S. BUTLER.

The following is Mr. Darwin's answer:-

January 3, 1880.

My Dear Sir, Dr. Krause, soon after the appearance of his article in Kosmos told me that he intended to publish it separately and to alter it considerably, and the altered MS. was sent to Mr. Dallas for translation. This is so common a practice that it never occurred to me to state that the article had been modified; but now I much regret that I did not do so. The original will soon appear in German, and I believe will be a much larger book than the English one; for, with Dr. Krause's consent, many long extracts from Miss Seward were omitted (as well as much other matter), from being in my opinion superfluous for the English reader. I believe that the omitted parts will appear as notes in the German edition. Should there be a reprint of the English Life I will state that the original as it appeared in Kosmos was modified by Dr. Krause before it was translated. I may add that I had obtained Dr. Krause's consent for a translation, and had arranged with Mr. Dallas before your book was announced. I remember this because Mr. Dallas wrote to tell me of the advertisement.--I remain, yours faithfully,

C. DARWIN."

This was not a letter I could accept. If Mr. Darwin had said that by some inadvertence, which he was unable to excuse or account for, a blunder had been made which he would at once correct so far as was in his power by a letter to the Times or the Athenaeum, and that a notice of the erratum should be printed on a flyleaf and pasted into all unsold copies of the "Life of Erasmus Darwin," there would have been no more heard about the matter from me; but when Mr. Darwin maintained that it was a common practice to take advantage of an opportunity of revising a work to interpolate a covert attack upon an opponent, and at the same time to misdate the interpolated matter by expressly stating that it appeared months sooner than it actually did, and prior to the work which it attacked; when he maintained that what was being done was "so common a practice that it never occurred," to him--the writer of some twenty volumes--to do what all literary men must know to be inexorably requisite, I thought this was going far beyond what was permissible in honourable warfare, and that it was time, in the interests of literary and scientific morality, even more than in my own, to appeal to public opinion. I was particularly struck with the use of the words "it never occurred to me," and felt how completely of a piece it was with the opening paragraph of the "Origin of Species." It was not merely that it did not occur to Mr. Darwin to state that the article had been modified since it was written--this would have been bad enough under the circumstances but that it did occur to him to go out of his way to say what was not true. There was no necessity for him to have said anything about my book. It appeared, moreover, inadequate to tell me that if a reprint of the English Life was wanted (which might or might not be the case, and if it was not the case, why, a shrug of the shoulders, and I must make the best of it), Mr. Darwin might perhaps silently omit his note about my book, as he omitted his misrepresentation of the author of the "Vestiges of Creation," and put the words "revised and corrected by the author" on his title-page.

No matter how high a writer may stand, nor what services he may have unquestionably rendered, it cannot be for the general well-being that he should be allowed to set aside the fundamental principles of straightforwardness and fair play. When I thought of Buffon, of Dr. Erasmus Darwin, of Lamarck and even of the author of the "Vestiges of Creation," to all of whom Mr. Darwin had dealt the same measure which

he was now dealing to myself; when I thought of these great men, now
dumb, who had borne the burden and heat of the day, and whose laurels
had been filched from them; of the manner, too, in which Mr. Darwin
had been abetted by those who should have been the first to detect
the fallacy which had misled him; of the hotbed of intrigue which
science has now become; of the disrepute into which we English must
fall as a nation if such practices as Mr. Darwin had attempted in
this case were to be tolerated;--when I thought of all this, I felt
that though prayers for the repose of dead men's souls might be
unavailing, yet a defence of their work and memory, no matter against
what odds, might avail the living, and resolved that I would do my
utmost to make my countrymen aware of the spirit now ruling among
those whom they delight to honour.

At first I thought I ought to continue the correspondence privately
with Mr. Darwin, and explain to him that his letter was insufficient,
but on reflection I felt that little good was likely to come of a
second letter, if what I had already written was not enough. I
therefore wrote to the Athenaeum and gave a condensed account of the
facts contained in the last ten or a dozen pages. My letter appeared
January 31, 1880. {50}

The accusation was a very grave one; it was made in a very public
place. I gave my name; I adduced the strongest prima facie grounds
for the acceptance of my statements; but there was no rejoinder, and
for the best of all reasons--that no rejoinder was possible.
Besides, what is the good of having a reputation for candour if one
may not stand upon it at a pinch? I never yet knew a person with an
especial reputation for candour without finding sooner or later that
he had developed it as animals develop their organs, through "sense
of need." Not only did Mr. Darwin remain perfectly quiet, but all
reviewers and litterateurs remained perfectly quiet also. It seemed-
-though I do not for a moment believe that this is so--as if public

opinion rather approved of what Mr. Darwin had done, and of his silence than otherwise. I saw the "Life of Erasmus Darwin" more frequently and more prominently advertised now than I had seen it hitherto--perhaps in the hope of selling off the adulterated copies, and being able to reprint the work with a corrected title page. Presently I saw Professor Huxley hastening to the rescue with his lecture on the coming of age of the "Origin of Species," and by May it was easy for Professor Ray Lankester to imply that Mr. Darwin was the greatest of living men. I have since noticed two or three other controversies raging in the Athenaeum and Times; in each of these cases I saw it assumed that the defeated party, when proved to have publicly misrepresented his adversary, should do his best to correct in public the injury which he had publicly inflicted, but I noticed that in none of them had the beaten side any especial reputation for candour. This probably made all the difference. But however this may be, Mr. Darwin left me in possession of the field, in the hope, doubtless, that the matter would blow over--which it apparently soon did. Whether it has done so in reality or no, is a matter which remains to be seen. My own belief is that people paid no attention to what I said, as believing it simply incredible, and that when they come to know that it is true, they will think as I do concerning it.

From ladies and gentlemen of science I admit that I have no expectations. There is no conduct so dishonourable that people will not deny it or explain it away, if it has been committed by one whom they recognise as of their own persuasion. It must be remembered that facts cannot be respected by the scientist in the same way as by other people. It is his business to familiarise himself with facts, and, as we all know, the path from familiarity to contempt is an easy one.

Here, then, I take leave of this matter for the present. If it appears that I have used language such as is rarely seen in controversy, let the reader remember that the occasion is, so far as I know, unparalleled for the cynicism and audacity with which the wrong complained of was committed and persisted in. I trust, however, that, though not indifferent to this, my indignation has been mainly roused, as when I wrote "Evolution, Old and New," before Mr. Darwin had given me personal ground of complaint against him, by the wrongs he has inflicted on dead men, on whose behalf I now fight, as I trust that some one--whom I thank by anticipation--may one day fight on mine.

CHAPTER V

Introduction to Professor Hering's lecture.

After I had finished "Evolution, Old and New," I wrote some articles for the Examiner, {52} in which I carried out the idea put forward in "Life and Habit," that we are one person with our ancestors. It follows from this, that all living animals and vegetables, being--as appears likely if the theory of evolution is accepted--descended from a common ancestor, are in reality one person, and unite to form a body corporate, of whose existence, however, they are unconscious. There is an obvious analogy between this and the manner in which the component cells of our bodies unite to form our single individuality, of which it is not likely they have a conception, and with which they have probably only the same partial and imperfect sympathy as we, the body corporate, have with them. In the articles above alluded to I separated the organic from the inorganic, and when I came to rewrite them, I found that this could not be done, and that I must reconstruct what I had written. I was at work on this--to which I

hope to return shortly--when Dr. Krause's' "Erasmus Darwin," with its
preliminary notice by Mr. Charles Darwin, came out, and having been
compelled, as I have shown above, by Dr. Krause's work to look a
little into the German language, the opportunity seemed favourable
for going on with it and becoming acquainted with Professor Hering's
lecture. I therefore began to translate his lecture at once, with
the kind assistance of friends whose patience seemed inexhaustible,
and found myself well rewarded for my trouble.

Professor Hering and I, to use a metaphor of his own, are as men who
have observed the action of living beings upon the stage of the
world, he from the point of view at once of a spectator and of one
who has free access to much of what goes on behind the scenes, I from
that of a spectator only, with none but the vaguest notion of the
actual manner in which the stage machinery is worked. If two men so
placed, after years of reflection, arrive independently of one
another at an identical conclusion as regards the manner in which
this machinery must have been invented and perfected, it is natural
that each should take a deep interest in the arguments of the other,
and be anxious to put them forward with the utmost possible
prominence. It seems to me that the theory which Professor Hering
and I are supporting in common, is one the importance of which is
hardly inferior to that of the theory of evolution itself--for it
puts the backbone, as it were, into the theory of evolution. I shall
therefore make no apology for laying my translation of Professor
Hering's work before my reader.

Concerning the identity of the main idea put forward in "Life and
Habit" with that of Professor Hering's lecture, there can hardly, I
think, be two opinions. We both of us maintain that we grow our
limbs as we do, and possess the instincts we possess, because we
remember having grown our limbs in this way, and having had these
instincts in past generations when we were in the persons of our

forefathers--each individual life adding a small (but so small, in any one lifetime, as to be hardly appreciable) amount of new experience to the general store of memory; that we have thus got into certain habits which we can now rarely break; and that we do much of what we do unconsciously on the same principle as that (whatever it is) on which we do all other habitual actions, with the greater ease and unconsciousness the more often we repeat them. Not only is the main idea the same, but I was surprised to find how often Professor Hering and I had taken the same illustrations with which to point our meaning.

Nevertheless, we have each of us left undealt with some points which the other has treated of. Professor Hering, for example, goes into the question of what memory is, and this I did not venture to do. I confined myself to saying that whatever memory was, heredity was also. Professor Hering adds that memory is due to vibrations of the molecules of the nerve fibres, which under certain circumstances recur, and bring about a corresponding recurrence of visible action.

This approaches closely to the theory concerning the physics of memory which has been most generally adopted since the time of Bonnet, who wrote as follows:-

"The soul never has a new sensation but by the inter position of the senses. This sensation has been originally attached to the motion of certain fibres. Its reproduction or recollection by the senses will then be likewise connected with these same fibres." . . . {54a}

And again:-

"It appeared to me that since this memory is connected with the body, it must depend upon some change which must happen to the primitive state of the sensible fibres by the action of objects. I have, therefore, admitted as probable that the state of the fibres on which an object has acted is not precisely the same after this action as it was before I have conjectured that the sensible fibres experience more or less durable modifications, which constitute the physics of memory and recollection." {54b}

Professor Hering comes near to endorsing this view, and uses it for the purpose of explaining personal identity. This, at least, is what he does in fact, though perhaps hardly in words. I did not say more upon the essence of personality than that it was inseparable from the idea that the various phases of our existence should have flowed one out of the other, "in what we see as a continuous, though it may be at times a very troubled, stream" {55} but I maintained that the identity between two successive generations was of essentially the same kind as that existing between an infant and an octogenarian. I thus left personal identity unexplained, though insisting that it was the key to two apparently distinct sets of phenomena, the one of which had been hitherto considered incompatible with our ideas concerning it. Professor Hering insists on this too, but he gives us farther insight into what personal identity is, and explains how it is that the phenomena of heredity are phenomena also of personal identity.

He implies, though in the short space at his command he has hardly said so in express terms, that personal identity as we commonly think of it--that is to say, as confined to the single life of the individual--consists in the uninterruptedness of a sufficient number of vibrations, which have been communicated from molecule to molecule of the nerve fibres, and which go on communicating each one of them its own peculiar characteristic elements to the new matter which we introduce into the body by way of nutrition. These vibrations may be so gentle as to be imperceptible for years together; but they are there, and may become perceived if they receive accession through the running into them of a wave going the same way as themselves, which wave has been set up in the ether by exterior objects and has been communicated to the organs of sense.

As these pages are on the point of leaving my hands, I see the following remarkable passage in Mind for the current month, and introduce it parenthetically here:-

"I followed the sluggish current of hyaline material issuing from globules of most primitive living substance. Persistently it followed its way into space, conquering, at first, the manifold resistances opposed to it by its watery medium. Gradually, however, its energies became exhausted, till at last, completely overwhelmed, it stopped, an immovable projection stagnated to death-like rigidity. Thus for hours, perhaps, it remained stationary, one of many such rays of some of the many kinds of protoplasmic stars. By degrees, then, or perhaps quite suddenly, HELP WOULD COME TO IT FROM FOREIGN BUT CONGRUOUS SOURCES. IT WOULD SEEM TO COMBINE WITH OUTSIDE COMPLEMENTAL MATTER drifted to it at random. Slowly it would regain thereby its vital mobility. Shrinking at first, but gradually completely restored and reincorporated into the onward tide of life, it was ready to take part again in the progressive flow of a new ray." {56}

To return to the end of the last paragraph but one. If this is so--

but I should warn the reader that Professor Hering is not responsible for this suggestion, though it seems to follow so naturally from what he has said that I imagine he intended the inference to be drawn,--if this is so, assimilation is nothing else than the communication of its own rhythms from the assimilating to the assimilated substance, to the effacement of the vibrations or rhythms heretofore existing in this last; and suitability for food will depend upon whether the rhythms of the substance eaten are such as to flow harmoniously into and chime in with those of the body which has eaten it, or whether they will refuse to act in concert with the new rhythms with which they have become associated, and will persist obstinately in pursuing their own course. In this case they will either be turned out of the body at once, or will disconcert its arrangements, with perhaps fatal consequences. This comes round to the conclusion I arrived at in "Life and Habit," that assimilation was nothing but the imbuing of one thing with the memories of another. (See "Life and Habit," pp. 136, 137, 140, &c.)

It will be noted that, as I resolved the phenomena of heredity into phenomena of personal identity, and left the matter there, so Professor Hering resolves the phenomena of personal identity into the phenomena of a living mechanism whose equilibrium is disturbed by vibrations of a certain character--and leaves it there. We now want to understand more about the vibrations.

But if, according to Professor Hering, the personal identity of the single life consists in the uninterruptedness of vibrations, so also do the phenomena of heredity. For not only may vibrations of a certain violence or character be persistent unperceived for many years in a living body, and communicate themselves to the matter it has assimilated, but they may, and will, under certain circumstances, extend to the particle which is about to leave the parent body as the germ of its future offspring. In this minute piece of matter there must, if Professor Hering is right, be an infinity of rhythmic undulations incessantly vibrating with more or less activity, and ready to be set in more active agitation at a moment's warning, under due accession of vibration from exterior objects. On the occurrence of such stimulus, that is to say, when a vibration of a suitable rhythm from without concurs with one within the body so as to augment it, the agitation may gather such strength that the touch, as it were, is given to a house of cards, and the whole comes toppling over. This toppling over is what we call action; and when it is the result of the disturbance of certain usual arrangements in certain usual ways, we call it the habitual development and instinctive characteristics of the race. In either case, then, whether we consider the continued identity of the individual in what we call his single life, or those features in his offspring which we refer to heredity, the same explanation of the phenomena is applicable. It follows from this as a matter of course, that the continuation of life or personal identity in the individual and the race are fundamentally of the same kind, or, in other words, that there is a veritable prolongation of identity or oneness of personality between parents and offspring. Professor Hering reaches his conclusion by physical methods, while I reached mine, as I am told, by metaphysical. I never yet could understand what "metaphysics" and "metaphysical" mean; but I should have said I reached it by the exercise of a little common sense while regarding certain facts which are open to every one. There is, however, so far as I can see, no difference in the conclusion come to.

The view which connects memory with vibrations may tend to throw light upon that difficult question, the manner in which neuter bees acquire structures and instincts, not one of which was possessed by any of their direct ancestors. Those who have read "Life and Habit" may remember, I suggested that the food prepared in the stomachs of the nurse-bees, with which the neuter working bees are fed, might thus acquire a quasi-seminal character, and be made a means of communicating the instincts and structures in question. {58} If assimilation be regarded as the receiving by one substance of the rhythms or undulations from another, the explanation just referred to receives an accession of probability.

If it is objected that Professor Hering's theory as to continuity of vibrations being the key to memory and heredity involves the action of more wheels within wheels than our imagination can come near to comprehending, and also that it supposes this complexity of action as going on within a compass which no unaided eye can detect by reason of its littleness, so that we are carried into a fairy land with which sober people should have nothing to do, it may be answered that the case of light affords us an example of our being truly aware of a multitude of minute actions, the hundred million millionth part of which we should have declared to be beyond our ken, could we not incontestably prove that we notice and count them all with a very sufficient and creditable accuracy.

"Who would not," {59a} says Sir John Herschel, "ask for demonstration when told that a gnat's wing, in its ordinary flight, beats many hundred times in a second? or that there exist animated and regularly organised beings many thousands of whose bodies laid close together would not extend to an inch? But what are these to the astonishing truths which modern optical inquiries have disclosed, which teach us that every point of a medium through which a ray of light passes is affected with a succession of periodical movements, recurring regularly at equal intervals, no less than five hundred millions of millions of times in a second; that it is by such movements communicated to the nerves of our eyes that we see; nay, more, that it is the DIFFERENCE in the frequency of their recurrence which affects us with the sense of the diversity of colour; that, for instance, in acquiring the sensation of redness, our eyes are affected four hundred and eighty-two millions of millions of times; of yellowness, five hundred and forty-two millions of millions of times; and of violet, seven hundred and seven millions of millions of times per second? {59b} Do not such things sound more like the ravings of madmen than the sober conclusions of people in their waking senses? They are, nevertheless, conclusions to which any one may most certainly arrive who will only be at the pains of examining the chain of reasoning by which they have been obtained."

A man counting as hard as he can repeat numbers one after another, and never counting more than a hundred, so that he shall have no long words to repeat, may perhaps count ten thousand, or a hundred a hundred times over, in an hour. At this rate, counting night and day, and allowing no time for rest or refreshment, he would count one million in four days and four hours, or say four days only. To count a million a million times over, he would require four million days, or roughly ten thousand years; for five hundred millions of millions, he must have the utterly unrealisable period of five million years. Yet he actually goes through this stupendous piece of reckoning unconsciously hour after hour, day after day, it may be for eighty years, OFTEN IN EACH SECOND of daylight; and how much more by artificial or subdued light I do not know. He knows whether his eye is being struck five hundred millions of millions of times, or only four hundred and eighty-two millions of millions of times. He thus shows that he estimates or counts each set of vibrations, and registers them according to his results. If a man writes upon the back of a British Museum blotting-pad of the common nonpareil pattern, on which there are some thousands of small spaces each differing in colour from that which is immediately next to it, his eye will, nevertheless, without an effort assign its true colour to each one of these spaces. This implies that he is all the time counting and taking tally of the difference in the numbers of the vibrations from each one of the small spaces in question. Yet the

mind that is capable of such stupendous computations as these so long
as it knows nothing about them, makes no little fuss about the
conscious adding together of such almost inconceivably minute numbers
as, we will say, 2730169 and 5790135--or, if these be considered too
large, as 27 and 19. Let the reader remember that he cannot by any
effort bring before his mind the units, not in ones, BUT IN MILLIONS
OF MILLIONS of the processes which his visual organs are undergoing
second after second from dawn till dark, and then let him demur if he
will to the possibility of the existence in a germ, of currents and
undercurrents, and rhythms and counter rhythms, also by the million
of millions--each one of which, on being overtaken by the rhythm from
without that chimes in with and stimulates it, may be the beginning
of that unsettlement of equilibrium which results in the crash of
action, unless it is timely counteracted.

If another objector maintains that the vibrations within the germ as
above supposed must be continually crossing and interfering with one
another in such a manner as to destroy the continuity of any one
series, it may be replied that the vibrations of the light proceeding
from the objects that surround us traverse one another by the
millions of millions every second yet in no way interfere with one
another. Nevertheless, it must be admitted that the difficulties of
the theory towards which I suppose Professor Hering to incline are
like those of all other theories on the same subject--almost
inconceivably great.

In "Life and Habit" I did not touch upon these vibrations, knowing
nothing about them. Here, then, is one important point of
difference, not between the conclusions arrived at, but between the
aim and scope of the work that Professor Hering and I severally
attempted. Another difference consists in the points at which we
have left off. Professor Hering, having established his main thesis,
is content. I, on the other hand, went on to maintain that if vigour

was due to memory, want of vigour was due to want of memory. Thus I was led to connect memory with the phenomena of hybridism and of old age; to show that the sterility of certain animals under domestication is only a phase of, and of a piece with, the very common sterility of hybrids--phenomena which at first sight have no connection either with each other or with memory, but the connection between which will never be lost sight of by those who have once laid hold of it. I also pointed out how exactly the phenomena of development agreed with those of the abeyance and recurrence of memory, and the rationale of the fact that puberty in so many animals and plants comes about the end of development. The principle underlying longevity follows as a matter of course. I have no idea how far Professor Hering would agree with me in the position I have taken in respect of these phenomena, but there is nothing in the above at variance with his lecture.

Another matter on which Professor Hering has not touched is the bearing of his theory on that view of evolution which is now commonly accepted. It is plain he accepts evolution, but it does not appear that he sees how fatal his theory is to any view of evolution except a teleological one--the purpose residing within the animal and not without it. There is, however, nothing in his lecture to indicate that he does not see this.

It should be remembered that the question whether memory is due to the persistence within the body of certain vibrations, which have been already set up within the bodies of its ancestors, is true or no, will not affect the position I took up in "Life and Habit." In that book I have maintained nothing more than that whatever memory is heredity is also. I am not committed to the vibration theory of memory, though inclined to accept it on a prima facie view. All I am committed to is, that if memory is due to persistence of vibrations, so is heredity; and if memory is not so due, then no more is heredity.

Finally, I may say that Professor Hering's lecture, the passage quoted from Dr. Erasmus Darwin on p. 26 of this volume, and a few hints in the extracts from Mr. Patrick Mathew which I have quoted in "Evolution, Old and New," are all that I yet know of in other writers as pointing to the conclusion that the phenomena of heredity are phenomena also of memory.

CHAPTER VI

Professor Ewald Hering "On Memory."

I will now lay before the reader a translation of Professor Hering's own words. I have had it carefully revised throughout by a gentleman whose native language is German, but who has resided in England for many years past. The original lecture is entitled "On Memory as a Universal Function of Organised Matter," and was delivered at the anniversary meeting of the Imperial Academy of Sciences at Vienna, May 30, 1870. {63} It is as follows:-

"When the student of Nature quits the narrow workshop of his own particular inquiry, and sets out upon an excursion into the vast kingdom of philosophical investigation, he does so, doubtless, in the hope of finding the answer to that great riddle, to the solution of a small part of which he devotes his life. Those, however, whom he leaves behind him still working at their own special branch of inquiry, regard his departure with secret misgivings on his behalf, while the born citizens of the kingdom of speculation among whom he would naturalise himself, receive him with well-authorised distrust. He is likely, therefore, to lose ground with the first, while not gaining it with the second.

The subject to the consideration of which I would now solicit your attention does certainly appear likely to lure us on towards the flattering land of speculation, but bearing in mind what I have just said, I will beware of quitting the department of natural science to which I have devoted myself hitherto. I shall, however, endeavour to attain its highest point, so as to take a freer view of the surrounding territory.

It will soon appear that I should fail in this purpose if my remarks were to confine themselves solely to physiology. I hope to show how far psychological investigations also afford not only permissible, but indispensable, aid to physiological inquiries.

Consciousness is an accompaniment of that animal and human organisation and of that material mechanism which it is the province of physiology to explore; and as long as the atoms of the brain follow their due course according to certain definite laws, there arises an inner life which springs from sensation and idea, from feeling and will.

We feel this in our own cases; it strikes us in our converse with other people; we can see it plainly in the more highly organised animals; even the lowest forms of life bear traces of it; and who can draw a line in the kingdom of organic life, and say that it is here the soul ceases?

With what eyes, then, is physiology to regard this two-fold life of the organised world? Shall she close them entirely to one whole side of it, that she may fix them more intently on the other?

So long as the physiologist is content to be a physicist, and nothing more--using the word "physicist" in its widest signification--his position in regard to the organic world is one of extreme but legitimate one-sidedness. As the crystal to the mineralogist or the vibrating string to the acoustician, so from this point of view both man and the lower animals are to the physiologist neither more nor less than the matter of which they consist. That animals feel desire and repugnance, that the material mechanism of the human frame is in chose connection with emotions of pleasure or pain, and with the active idea-life of consciousness--this cannot, in the eyes of the physicist, make the animal or human body into anything more than what it actually is. To him it is a combination of matter, subjected to the same inflexible laws as stones and plants--a material combination, the outward and inward movements of which interact as cause and effect, and are in as close connection with each other and with their surroundings as the working of a machine with the revolutions of the wheels that compose it.

Neither sensation, nor idea, nor yet conscious will, can form a link in this chain of material occurrences which make up the physical life of an organism. If I am asked a question and reply to it, the material process which the nerve fibre conveys from the organ of hearing to the brain must travel through my brain as an actual and material process before it can reach the nerves which will act upon my organs of speech. It cannot, on reaching a given place in the brain, change then and there into an immaterial something, and turn up again some time afterwards in another part of the brain as a material process. The traveller in the desert might as well hope, before he again goes forth into the wilderness of reality, to take rest and refreshment in the oasis with which the Fata Morgana illudes him; or as well might a prisoner hope to escape from his prison through a door reflected in a mirror.

So much for the physiologist in his capacity of pure physicist. As
long as he remains behind the scenes in painful exploration of the
details of the machinery--as long as he only observes the action of
the players from behind the stage--so long will he miss the spirit of
the performance, which is, nevertheless, caught easily by one who
sees it from the front. May he not, then, for once in a way, be
allowed to change his standpoint? True, he came not to see the
representation of an imaginary world; he is in search of the actual;
but surely it must help him to a comprehension of the dramatic
apparatus itself, and of the manner in which it is worked, if he were
to view its action from in front as well as from behind, or at least
allow himself to hear what sober-minded spectators can tell him upon
the subject.

There can be no question as to the answer; and hence it comes that
psychology is such an indispensable help to physiology, whose fault
it only in small part is that she has hitherto made such little use
of this assistance; for psychology has been late in beginning to till
her fertile field with the plough of the inductive method, and it is
only from ground so tilled that fruits can spring which can be of
service to physiology.

If, then, the student of nervous physiology takes his stand between
the physicist and the psychologist, and if the first of these rightly
makes the unbroken causative continuity of all material processes an
axiom of his system of investigation, the prudent psychologist, on
the other hand, will investigate the laws of conscious life according
to the inductive method, and will hence, as much as the physicist,
make the existence of fixed laws his initial assumption. If, again,
the most superficial introspection teaches the physiologist that his
conscious life is dependent upon the mechanical adjustments of his
body, and that inversely his body is subjected with certain

limitations to his will, then it only remains for him to make one
assumption more, namely, THAT THIS MUTUAL INTERDEPENDENCE BE-
TWEEN THE
SPIRITUAL AND THE MATERIAL IS ITSELF ALSO DEPENDENT ON LAW,
and he
has discovered the bond by which the science of matter and the
science of consciousness are united into a single whole.

Thus regarded, the phenomena of consciousness become functions of the
material changes of organised substance, and inversely--though this
is involved in the use of the word "function"--the material processes
of brain substance become functions of the phenomena of
consciousness. For when two variables are so dependent upon one
another in the changes they undergo in accordance with fixed laws
that a change in either involves simultaneous and corresponding
change in the other, the one is called a function of the other.

This, then, by no means implies that the two variables above-named--
matter and consciousness--stand in the relation of cause and effect,
antecedent and consequence, to one another. For on this subject we
know nothing.

The materialist regards consciousness as a product or result of
matter, while the idealist holds matter to be a result of
consciousness, and a third maintains that matter and spirit are
identical; with all this the physiologist, as such, has nothing
whatever to do; his sole concern is with the fact that matter and
consciousness are functions one of the other.

By the help of this hypothesis of the functional interdependence of
matter and spirit, modern physiology is enabled to bring the
phenomena of consciousness within the domain of her investigations
without leaving the terra firma of scientific methods. The

physiologist, as physicist, can follow the ray of light and the wave of sound or heat till they reach the organ of sense. He can watch them entering upon the ends of the nerves, and finding their way to the cells of the brain by means of the series of undulations or vibrations which they establish in the nerve filaments. Here, however, he loses all trace of them. On the other hand, still looking with the eyes of a pure physicist, he sees sound waves of speech issue from the mouth of a speaker; he observes the motion of his own limbs, and finds how this is conditional upon muscular contractions occasioned by the motor nerves, and how these nerves are in their turn excited by the cells of the central organ. But here again his knowledge comes to an end. True, he sees indications of the bridge which is to carry him from excitation of the sensory to that of the motor nerves in the labyrinth of intricately interwoven nerve cells, but he knows nothing of the inconceivably complex process which is introduced at this stage. Here the physiologist will change his standpoint; what matter will not reveal to his inquiry, he will find in the mirror, as it were, of consciousness; by way of a reflection, indeed, only, but a reflection, nevertheless, which stands in intimate relation to the object of his inquiry. When at this point he observes how one idea gives rise to another, how closely idea is connected with sensation and sensation with will, and how thought, again, and feeling are inseparable from one another, he will be compelled to suppose corresponding successions of material processes, which generate and are closely connected with one another, and which attend the whole machinery of conscious life, according to the law of the functional interdependence of matter and consciousness.

After this explanation I shall venture to regard under a single aspect a great series of phenomena which apparently have nothing to do with one another, and which belong partly to the conscious and

partly to the unconscious life of organised beings. I shall regard
them as the outcome of one and the same primary force of organised
matter--namely, its memory or power of reproduction.

The word "memory" is often understood as though it meant nothing more
than our faculty of intentionally reproducing ideas or series of
ideas. But when the figures and events of bygone days rise up again
unbidden in our minds, is not this also an act of recollection or
memory? We have a perfect right to extend our conception of memory
so as to make it embrace involuntary reproductions, of sensations,
ideas, perceptions, and efforts; but we find, on having done so, that
we have so far enlarged her boundaries that she proves to be an
ultimate and original power, the source, and at the same time the
unifying bond, of our whole conscious life.

We know that when an impression, or a series of impressions, has been
made upon our senses for a long time, and always in the same way, it
may come to impress itself in such a manner upon the so-called sense-
memory that hours afterwards, and though a hundred other things have
occupied our attention meanwhile, it will yet return suddenly to our
consciousness with all the force and freshness of the original
sensation. A whole group of sensations is sometimes reproduced in
its due sequence as regards time and space, with so much reality that
it illudes us, as though things were actually present which have long
ceased to be so. We have here a striking proof of the fact that
after both conscious sensation and perception have been extinguished,
their material vestiges yet remain in our nervous system by way of a
change in its molecular or atomic disposition, {69} that enables the
nerve substance to reproduce all the physical processes of the
original sensation, and with these the corresponding psychical
processes of sensation and perception.

Every hour the phenomena of sense-memory are present with each one of us, but in a less degree than this. We are all at times aware of a host of more or less faded recollections of earlier impressions, which we either summon intentionally or which come upon us involuntarily. Visions of absent people come and go before us as faint and fleeting shadows, and the notes of long-forgotten melodies float around us, not actually heard, but yet perceptible.

Some things and occurrences, especially if they have happened to us only once and hurriedly, will be reproducible by the memory in respect only of a few conspicuous qualities; in other cases those details alone will recur to us which we have met with elsewhere, and for the reception of which the brain is, so to speak, attuned. These last recollections find themselves in fuller accord with our consciousness, and enter upon it more easily and energetically; hence also their aptitude for reproduction is enhanced; so that what is common to many things, and is therefore felt and perceived with exceptional frequency, becomes reproduced so easily that eventually the actual presence of the corresponding external stimuli is no longer necessary, and it will recur on the vibrations set up by faint stimuli from within. {70} Sensations arising in this way from within, as, for example, an idea of whiteness, are not, indeed, perceived with the full freshness of those raised by the actual presence of white light without us, but they are of the same kind; they are feeble repetitions of one and the same material brain process--of one and the same conscious sensation. Thus the idea of whiteness arises in our mind as a faint, almost extinct, sensation.

In this way those qualities which are common to many things become separated, as it were, in our memory from the objects with which they were originally associated, and attain an independent existence in our consciousness as IDEAS and CONCEPTIONS, and thus the whole rich superstructure of our ideas and conceptions is built up from materials supplied by memory.

On examining more closely, we see plainly that memory is a faculty
not only of our conscious states, but also, and much more so, of our
unconscious ones. I was conscious of this or that yesterday, and am
again conscious of it to-day. Where has it been meanwhile? It does
not remain continuously within my consciousness, nevertheless it
returns after having quitted it. Our ideas tread but for a moment
upon the stage of consciousness, and then go back again behind the
scenes, to make way for others in their place. As the player is only
a king when he is on the stage, so they too exist as ideas so long
only as they are recognised. How do they live when they are off the
stage? For we know that they are living somewhere; give them their
cue and they reappear immediately. They do not exist continuously as
ideas; what is continuous is the special disposition of nerve
substance in virtue of which this substance gives out to-day the same
sound which it gave yesterday if it is rightly struck. {71}
Countless reproductions of organic processes of our brain connect
themselves orderly together, so that one acts as a stimulus to the
next, but a phenomenon of consciousness is not necessarily attached
to every link in the chain. From this it arises that a series of
ideas may appear to disregard the order that would be observed in
purely material processes of brain substance unaccompanied by
consciousness; but on the other hand it becomes possible for a long
chain of recollections to have its due development without each link
in the chain being necessarily perceived by ourselves. One may
emerge from the bosom of our unconscious thoughts without fully
entering upon the stage of conscious perception; another dies away in
unconsciousness, leaving no successor to take its place. Between the
"me" of to-day and the "me" of yesterday lie night and sleep, abysses
of unconsciousness; nor is there any bridge but memory with which to
span them. Who can hope after this to disentangle the infinite
intricacy of our inner life? For we can only follow its threads so
far as they have strayed over within the bounds of consciousness. We

might as well hope to familiarise ourselves with the world of forms that teem within the bosom of the sea by observing the few that now and again come to the surface and soon return into the deep.

The bond of union, therefore, which connects the individual phenomena of our consciousness lies in our unconscious world; and as we know nothing of this but what investigation into the laws of matter teach us--as, in fact, for purely experimental purposes, "matter" and the "unconscious" must be one and the same thing--so the physiologist has a full right to denote memory as, in the wider sense of the word, a function of brain substance, whose results, it is true, fall, as regards one part of them, into the domain of consciousness, while another and not less essential part escapes unperceived as purely material processes.

The perception of a body in space is a very complicated process. I see suddenly before me, for example, a white ball. This has the effect of conveying to me more than a mere sensation of whiteness. I deduce the spherical character of the ball from the gradations of light and shade upon its surface. I form a correct appreciation of its distance from my eye, and hence again I deduce an inference as to the size of the ball. What an expenditure of sensations, ideas, and inferences is found to be necessary before all this can be brought about; yet the production of a correct perception of the ball was the work only of a few seconds, and I was unconscious of the individual processes by means of which it was effected, the result as a whole being alone present in my consciousness.

The nerve substance preserves faithfully the memory of habitual actions. {72} Perceptions which were once long and difficult, requiring constant and conscious attention, come to reproduce themselves in transient and abridged guise, without such duration and intensity that each link has to pass over the threshold of our consciousness.

We have chains of material nerve processes to which eventually a link becomes attached that is attended with conscious perception. This is sufficiently established from the standpoint of the physiologist, and is also proved by our unconsciousness of many whole series of ideas and of the inferences we draw from them. If the soul is not to ship through the fingers of physiology, she must hold fast to the considerations suggested by our unconscious states. As far, however, as the investigations of the pure physicist are concerned, the unconscious and matter are one and the same thing, and the physiology of the unconscious is no "philosophy of the unconscious."

By far the greater number of our movements are the result of long and arduous practice. The harmonious cooperation of the separate muscles, the finely adjusted measure of participation which each contributes to the working of the whole, must, as a rule, have been laboriously acquired, in respect of most of the movements that are necessary in order to effect it. How long does it not take each note to find its way from the eyes to the fingers of one who is beginning to learn the pianoforte; and, on the other hand, what an astonishing performance is the playing of the professional pianist. The sight of each note occasions the corresponding movement of the fingers with the speed of thought--a hurried glance at the page of music before him suffices to give rise to a whole series of harmonies; nay, when a melody has been long practised, it can be played even while the player's attention is being given to something of a perfectly different character over and above his music.

The will need now no longer wend its way to each individual finger before the desired movements can be extorted from it; no longer now does a sustained attention keep watch over the movements of each limb; the will need exercise a supervising control only. At the word of command the muscles become active, with a due regard to time and proportion, and go on working, so long as they are bidden to keep in their accustomed groove, while a slight hint on the part of the will, will indicate to them their further journey. How could all this be if every part of the central nerve system, by means of which movement is effected, were not able {74a} to reproduce whole series of vibrations, which at an earlier date required the constant and continuous participation of consciousness, but which are now set in motion automatically on a mere touch, as it were, from consciousness--if it were not able to reproduce them the more quickly and easily in proportion to the frequency of the repetitions--if, in fact, there was no power of recollecting earlier performances? Our perceptive faculties must have remained always at their lowest stage if we had been compelled to build up consciously every process from the details of the sensation-causing materials tendered to us by our senses; nor could our voluntary movements have got beyond the helplessness of the child, if the necessary impulses could only be imparted to every movement through effort of the will and conscious reproduction of all the corresponding ideas--if, in a word, the motor nerve system had not also its memory, {74b} though that memory is unperceived by ourselves. The power of this memory is what is called "the force of habit."

It seems, then, that we owe to memory almost all that we either have or are; that our ideas and conceptions are its work, and that our every perception, thought, and movement is derived from this source. Memory collects the countless phenomena of our existence into a single whole; and as our bodies would be scattered into the dust of their component atoms if they were not held together by the attraction of matter, so our consciousness would be broken up into as many fragments as we had lived seconds but for the binding and unifying force of memory.

We have already repeatedly seen that the reproductions of organic processes, brought about by means of the memory of the nervous system, enter but partly within the domain of consciousness, remaining unperceived in other and not less important respects. This is also confirmed by numerous facts in the life of that part of the nervous system which ministers almost exclusively to our unconscious life processes. For the memory of the so-called sympathetic ganglionic system is no less rich than that of the brain and spinal marrow, and a great part of the medical art consists in making wise use of the assistance thus afforded us.

To bring, however, this part of my observations to a close, I will take leave of the nervous system, and glance hurriedly at other phases of organised matter, where we meet with the same powers of reproduction, but in simpler guise.

Daily experience teaches us that a muscle becomes the stronger the more we use it. The muscular fibre, which in the first instance may have answered but feebly to the stimulus conducted to it by the motor nerve, does so with the greater energy the more often it is stimulated, provided, of course, that reasonable times are allowed for repose. After each individual action it becomes more capable, more disposed towards the same kind of work, and has a greater aptitude for repetition of the same organic processes. It gains also in weight, for it assimilates more matter than when constantly at rest. We have here, in its simplest form, and in a phase which comes home most closely to the comprehension of the physicist, the same power of reproduction which we encountered when we were dealing with nerve substance, but under such far more complicated conditions. And what is known thus certainly from muscle substance holds good with greater or less plainness for all our organs. More especially may we note the fact, that after increased use, alternated with times of repose, there accrues to the organ in all animal economy an increased power of execution with an increased power of assimilation and a gain in size.

This gain in size consists not only in the enlargement of the individual cells or fibres of which the organ is composed, but in the multiplication of their number; for when cells have grown to a certain size they give rise to others, which inherit more or less completely the qualities of those from which they came, and therefore appear to be repetitions of the same cell. This growth, and multiplication of cells is only a special phase of those manifold functions which characterise organised matter, and which consist not only in what goes on within the cell substance as alterations or undulatory movement of the molecular disposition, but also in that which becomes visible outside the cells as change of shape, enlargement, or subdivision. Reproduction of performance, therefore, manifests itself to us as reproduction of the cells themselves, as may be seen most plainly in the case of plants, whose chief work consists in growth, whereas with animal organism other faculties greatly preponderate.

Let us now take a brief survey of a class of facts in the case of which we may most abundantly observe the power of memory in organised matter. We have ample evidence of the fact that characteristics of an organism may descend to offspring which the organism did not inherit, but which it acquired owing to the special circumstances under which it lived; and that, in consequence, every organism imparts to the germ that issues from it a small heritage of acquisitions which it has added during its own lifetime to the gross inheritance of its race.

When we reflect that we are dealing with the heredity of acquired qualities which came to development in the most diverse parts of the parent organism, it must seem in a high degree mysterious how those parts can have any kind of influence upon a germ which develops itself in an entirely different place. Many mystical theories have been propounded for the elucidation of this question, but the following reflections may serve to bring the cause nearer to the comprehension of the physiologist.

The nerve substance, in spite of its thousandfold subdivision as cells and fibres, forms, nevertheless, a united whole, which is present directly in all organs--nay, as more recent histology conjectures, in each cell of the more important organs--or is at least in ready communication with them by means of the living, irritable, and therefore highly conductive substance of other cells. Through the connection thus established all organs find themselves in such a condition of more or less mutual interdependence upon one another, that events which happen to one are repeated in others, and a notification, however slight, of a vibration set up {77} in one quarter is at once conveyed even to the farthest parts of the body. With this easy and rapid intercourse between all parts is associated the more difficult communication that goes on by way of the circulation of sap or blood.

We see, further, that the process of the development of all germs that are marked out for independent existence causes a powerful reaction, even from the very beginning of that existence, on both the conscious and unconscious life of the whole organism. We may see this from the fact that the organ of reproduction stands in closer and more important relation to the remaining parts, and especially to the nervous system, than do the other organs; and, inversely, that both the perceived and unperceived events affecting the whole organism find a more marked response in the reproductive system than elsewhere.

We can now see with sufficient plainness in what the material connection is established between the acquired peculiarities of an organism, and the proclivity on the part of the germ in virtue of which it develops the special characteristics of its parent.

The microscope teaches us that no difference can be perceived between one germ and another; it cannot, however, be objected on this account that the determining cause of its ulterior development must be something immaterial, rather than the specific kind of its material constitution.

The curves and surfaces which the mathematician conceives, or finds conceivable, are more varied and infinite than the forms of animal life. Let us suppose an infinitely small segment to be taken from every possible curve; each one of these will appear as like every other as one germ is to another, yet the whole of every curve lies dormant, as it were, in each of them, and if the mathematician chooses to develop it, it will take the path indicated by the elements of each segment.

It is an error, therefore, to suppose that such fine distinctions as physiology must assume lie beyond the limits of what is conceivable by the human mind. An infinitely small change of position on the part of a point, or in the relations of the parts of a segment of a curve to one another, suffices to alter the law of its whole path, and so in like manner an infinitely small influence exercised by the parent organism on the molecular disposition of the germ {78} may suffice to produce a determining effect upon its whole farther development.

What is the descent of special peculiarities but a reproduction on the part of organised matter of processes in which it once took part as a germ in the germ-containing organs of its parent, and of which it seems still to retain a recollection that reappears when time and

the occasion serve, inasmuch as it responds to the same or like
stimuli in a like way to that in which the parent organism responded,
of which it was once part, and in the events of whose history it was
itself also an accomplice? {79} When an action through long habit or
continual practice has become so much a second nature to any
organisation that its effects will penetrate, though ever so faintly,
into the germ that lies within it, and when this last comes to find
itself in a new sphere, to extend itself, and develop into a new
creature--(the individual parts of which are still always the
creature itself and flesh of its flesh, so that what is reproduced is
the same being as that in company with which the germ once lived, and
of which it was once actually a part)--all this is as wonderful as
when a grey-haired man remembers the events of his own childhood; but
it is not more so. Whether we say that the same organised substance
is again reproducing its past experience, or whether we prefer to
hold that an offshoot or part of the original substance has waxed and
developed itself since separation from the parent stock, it is plain
that this will constitute a difference of degree, not kind.

When we reflect upon the fact that unimportant acquired
characteristics can be reproduced in offspring, we are apt to forget
that offspring is only a full-sized reproduction of the parent--a
reproduction, moreover, that goes as far as possible into detail. We
are so accustomed to consider family resemblance a matter of course,
that we are sometimes surprised when a child is in some respect
unlike its parent; surely, however, the infinite number of points in
respect of which parents and children resemble one another is a more
reasonable ground for our surprise.

But if the substance of the germ can reproduce characteristics
acquired by the parent during its single life, how much more will it
not be able to reproduce those that were congenital to the parent,
and which have happened through countless generations to the

organised matter of which the germ of to-day is a fragment? We
cannot wonder that action already taken on innumerable past occasions
by organised matter is more deeply impressed upon the recollection of
the germ to which it gives rise than action taken once only during a
single lifetime. {80a}

We must bear in mind that every organised being now in existence
represents the last link of an inconceivably long series of
organisms, which come down in a direct line of descent, and of which
each has inherited a part of the acquired characteristics of its
predecessor. Everything, furthermore, points in the direction of our
believing that at the beginning of this chain there existed an
organism of the very simplest kind, something, in fact, like those
which we call organised germs. The chain of living beings thus
appears to be the magnificent achievement of the reproductive power
of the original organic structure from which they have all descended.
As this subdivided itself and transmitted its characteristics {80b}
to its descendants, these acquired new ones, and in their turn
transmitted them--all new germs transmitting the chief part of what
had happened to their predecessors, while the remaining part lapsed
out of their memory, circumstances not stimulating it to reproduce
itself.

An organised being, therefore, stands before us a product of the
unconscious memory of organised matter, which, ever increasing and
ever dividing itself, ever assimilating new matter and returning it
in changed shape to the inorganic world, ever receiving some new
thing into its memory, and transmitting its acquisitions by the way
of reproduction, grows continually richer and richer the longer it
lives.

Thus regarded, the development of one of the more highly organised
animals represents a continuous series of organised recollections

concerning the past development of the great chain of living forms, the last link of which stands before us in the particular animal we may be considering. As a complicated perception may arise by means of a rapid and superficial reproduction of long and laboriously practised brain processes, so a germ in the course of its development hurries through a series of phases, hinting at them only. Often and long foreshadowed in theories of varied characters, this conception has only now found correct exposition from a naturalist of our own time. {81} For Truth hides herself under many disguises from those who seek her, but in the end stands unveiled before the eyes of him whom she has chosen.

Not only is there a reproduction of form, outward and inner conformation of body, organs, and cells, but the habitual actions of the parent are also reproduced. The chicken on emerging from the eggshell runs off as its mother ran off before it; yet what an extraordinary complication of emotions and sensations is necessary in order to preserve equilibrium in running. Surely the supposition of an inborn capacity for the reproduction of these intricate actions can alone explain the facts. As habitual practice becomes a second nature to the individual during his single lifetime, so the often-repeated action of each generation becomes a second nature to the race.

The chicken not only displays great dexterity in the performance of movements for the effecting of which it has an innate capacity, but it exhibits also a tolerably high perceptive power. It immediately picks up any grain that may be thrown to it. Yet, in order to do this, more is wanted than a mere visual perception of the grains; there must be an accurate apprehension of the direction and distance of the precise spot in which each grain is lying, and there must be no less accuracy in the adjustment of the movements of the head and of the whole body. The chicken cannot have gained experience in

these respects while it was still in the egg. It gained it rather
from the thousands of thousands of beings that have lived before it,
and from which it is directly descended.

The memory of organised substance displays itself here in the most
surprising fashion. The gentle stimulus of the light proceeding from
the grain that affects the retina of the chicken, {82} gives occasion
for the reproduction of a many-linked chain of sensations,
perceptions, and emotions, which were never yet brought together in
the case of the individual before us. We are accustomed to regard
these surprising performances of animals as manifestations of what we
call instinct, and the mysticism of natural philosophy has ever shown
a predilection for this theme; but if we regard instinct as the
outcome of the memory or reproductive power of organised substance,
and if we ascribe a memory to the race as we already ascribe it to
the individual, then instinct becomes at once intelligible, and the
physiologist at the same time finds a point of contact which will
bring it into connection with the great series of facts indicated
above as phenomena of the reproductive faculty. Here, then, we have
a physical explanation which has not, indeed, been given yet, but the
time for which appears to be rapidly approaching.

When, in accordance with its instinct, the caterpillar becomes a
chrysalis, or the bird builds its nest, or the bee its cell, these
creatures act consciously and not as blind machines. They know how
to vary their proceedings within certain limits in conformity with
altered circumstances, and they are thus liable to make mistakes.
They feel pleasure when their work advances and pain if it is
hindered; they learn by the experience thus acquired, and build on a
second occasion better than on the first; but that even in the outset
they hit so readily upon the most judicious way of achieving their
purpose, and that their movements adapt themselves so admirably and
automatically to the end they have in view--surely this is owing to

the inherited acquisitions of the memory of their nerve substance, which requires but a touch and it will fall at once to the most appropriate kind of activity, thinking always, and directly, of whatever it is that may be wanted.

Man can readily acquire surprising kinds of dexterity if he confines his attention to their acquisition. Specialisation is the mother of proficiency. He who marvels at the skill with which the spider weaves her web should bear in mind that she did not learn her art all on a sudden, but that innumerable generations of spiders acquired it toilsomely and step by step--this being about all that, as a general rule, they did acquire. Man took to bows and arrows if his nets failed him--the spider starved. Thus we see the body and--what most concerns us--the whole nervous system of the new-born animal constructed beforehand, and, as it were, ready attuned for intercourse with the outside world in which it is about to play its part, by means of its tendency to respond to external stimuli in the same manner as it has often heretofore responded in the persons of its ancestors.

We naturally ask whether the brain and nervous system of the human infant are subjected to the principles we have laid down above? Man certainly finds it difficult to acquire arts of which the lower animals are born masters; but the brain of man at birth is much farther from its highest development than is the brain of an animal. It not only grows for a longer time, but it becomes stronger than that of other living beings. The brain of man may be said to be exceptionally young at birth. The lower animal is born precocious, and acts precociously; it resembles those infant prodigies whose brain, as it were, is born old into the world, but who, in spite of, or rather in addition to, their rich endowment at birth, in after life develop as much mental power as others who were less splendidly furnished to start with, but born with greater freshness of youth.

Man's brain, and indeed his whole body, affords greater scope for individuality, inasmuch as a relatively greater part of it is of post-natal growth. It develops under the influence of impressions made by the environment upon its senses, and thus makes its acquisitions in a more special and individual manner, whereas the animal receives them ready made, and of a more final, stereotyped character.

Nevertheless, it is plain we must ascribe both to the brain and body of the new-born infant a far-reaching power of remembering or reproducing things which have already come to their development thousands of times over in the persons of its ancestors. It is in virtue of this that it acquires proficiency in the actions necessary for its existence--so far as it was not already at birth proficient in them--much more quickly and easily than would be otherwise possible; but what we call instinct in the case of animals takes in man the looser form of aptitude, talent, and genius. {84} Granted that certain ideas are not innate, yet the fact of their taking form so easily and certainly from out of the chaos of his sensations, is due not to his own labour, but to that of the brain substance of the thousands of thousands of generations from whom he is descended. Theories concerning the development of individual consciousness which deny heredity or the power of transmission, and insist upon an entirely fresh start for every human soul, as though the infinite number of generations that have gone before us might as well have never lived for all the effect they have had upon ourselves,--such theories will contradict the facts of our daily experience at every touch and turn.

The brain processes and phenomena of consciousness which ennoble man in the eyes of his fellows have had a less ancient history than those connected with his physical needs. Hunger and the reproductive instinct affected the oldest and simplest forms of the organic world.

It is in respect of these instincts, therefore, and of the means to gratify them, that the memory of organised substance is strongest-- the impulses and instincts that arise hence having still paramount power over the minds of men. The spiritual life has been superadded slowly; its most splendid outcome belongs to the latest epoch in the history of organised matter, nor has any very great length of time elapsed since the nervous system was first crowned with the glory of a large and well-developed brain.

Oral tradition and written history have been called the memory of man, and this is not without its truth. But there is another and a living memory in the innate reproductive power of brain substance, and without this both writings and oral tradition would be without significance to posterity. The most sublime ideas, though never so immortalised in speech or letters, are yet nothing for heads that are out of harmony with them; they must be not only heard, but reproduced; and both speech and writing would be in vain were there not an inheritance of inward and outward brain development, growing in correspondence with the inheritance of ideas that are handed down from age to age, and did not an enhanced capacity for their reproduction on the part of each succeeding generation accompany the thoughts that have been preserved in writing. Man's conscious memory comes to an end at death, but the unconscious memory of Nature is true and ineradicable: whoever succeeds in stamping upon her the impress of his work, she will remember him to the end of time.

CHAPTER VII

Introduction to a translation of the chapter upon instinct in Von

Hartmann's "Philosophy of the Unconscious."

I am afraid my readers will find the chapter on instinct from Von Hartmann's "Philosophy of the Unconscious," which will now follow, as distasteful to read as I did to translate, and would gladly have spared it them if I could. At present, the works of Mr. Sully, who has treated of the "Philosophy of the Unconscious" both in the Westminster Review (vol. xlix. N.S.) and in his work "Pessimism," are the best source to which English readers can have recourse for information concerning Von Hartmann. Giving him all credit for the pains he has taken with an ungrateful, if not impossible subject, I think that a sufficient sample of Von Hartmann's own words will be a useful adjunct to Mr. Sully's work, and may perhaps save some readers trouble by resolving them to look no farther into the "Philosophy of the Unconscious." Over and above this, I have been so often told that the views concerning unconscious action contained in the foregoing lecture and in "Life and Habit" are only the very fallacy of Von Hartmann over again, that I should like to give the public an opportunity of seeing whether this is so or no, by placing the two contending theories of unconscious action side by side. I hope that it will thus be seen that neither Professor Hering nor I have fallen into the fallacy of Von Hartmann, but that rather Von Hartmann has fallen into his fallacy through failure to grasp the principle which Professor Hering has insisted upon, and to connect heredity with memory.

Professor Hering's philosophy of the unconscious is of extreme simplicity. He rests upon a fact of daily and hourly experience, namely, that practice makes things easy that were once difficult, and often results in their being done without any consciousness of effort. But if the repetition of an act tends ultimately, under certain circumstances, to its being done unconsciously, so also is the fact of an intricate and difficult action being done unconsciously an argument that it must have been done repeatedly already. As I said in "Life and Habit," it is more easy to suppose that occasions on which such an action has been performed have not been wanting, even though we do not see when and where they were, than that the facility which we observe should have been attained without practice and memory (p. 56).

There can be nothing better established or more easy, whether to understand or verify, than the unconsciousness with which habitual actions come to be performed. If, however, it is once conceded that it is the manner of habitual action generally, then all a priori objection to Professor Hering's philosophy of the unconscious is at an end. The question becomes one of fact in individual cases, and of degree.

How far, then, does the principle of the convertibility, as it were, of practice and unconsciousness extend? Can any line be drawn beyond which it shall cease to operate? If not, may it not have operated and be operating to a vast and hitherto unsuspected extent? This is all, and certainly it is sufficiently simple. I sometimes think it has found its greatest stumbling-block in its total want of mystery, as though we must be like those conjurers whose stock in trade is a small deal table and a kitchen-chair with bare legs, and who, with their parade of "no deception" and "examine everything for yourselves," deceive worse than others who make use of all manner of elaborate paraphernalia. It is true we require no paraphernalia, and we produce unexpected results, but we are not conjuring.

To turn now to Von Hartmann. When I read Mr. Sully's article in the Westminster Review, I did not know whether the sense of mystification which it produced in me was wholly due to Von Hartmann or no; but on making acquaintance with Von Hartmann himself, I found that Mr. Sully has erred, if at all, in making him more intelligible than he

actually is. Von Hartmann has not got a meaning. Give him Professor
Hering's key and he might get one, but it would be at the expense of
seeing what approach he had made to a system fallen to pieces.
Granted that in his details and subordinate passages he often both
has and conveys a meaning, there is, nevertheless, no coherence
between these details, and the nearest approach to a broad conception
covering the work which the reader can carry away with him is at once
so incomprehensible and repulsive, that it is difficult to write
about it without saying more perhaps than those who have not seen the
original will accept as likely to be true. The idea to which I refer
is that of an unconscious clairvoyance, which, from the language
continually used concerning it, must be of the nature of a person,
and which is supposed to take possession of living beings so fully as
to be the very essence of their nature, the promoter of their
embryonic development, and the instigator of their instinctive
actions. This approaches closely to the personal God of Mosaic and
Christian theology, with the exception that the word "clairvoyance"
{89} is substituted for God, and that the God is supposed to be
unconscious.

Mr. Sully says:-

"When we grasp it [the philosophy of Von Hartmann] as a whole, it
amounts to nothing more than this, that all or nearly all the
phenomena of the material and spiritual world rest upon and result
from a mysterious, unconscious being, though to call it being is
really to add on an idea not immediately contained within the all-
sufficient principle. But what difference is there between this and
saying that the phenomena of the world at large come we know not
whence? . . . The unconscious, therefore, tends to be simple phrase
and nothing more . . . No doubt there are a number of mental
processes . . . of which we are unconscious . . . but to infer from

this that they are due to an unconscious power, and to proceed to demonstrate them in the presence of the unconscious through all nature, is to make an unwarrantable saltus in reasoning. What, in fact, is this 'unconscious' but a high-sounding name to veil our ignorance? Is the unconscious any better explanation of phenomena we do not understand than the 'devil-devil' by which Australian tribes explain the Leyden jar and its phenomena? Does it increase our knowledge to know that we do not know the origin of language or the cause of instinct? . . . Alike in organic creation and the evolution of history 'performances and actions'--the words are those of Strauss--are ascribed to an unconscious, which can only belong to a conscious being. {90a}

. . . . .

"The difficulties of the system advance as we proceed. {90b} Subtract this questionable factor--the unconscious from Hartmann's 'Biology and Psychology,' and the chapters remain pleasant and instructive reading. But with the third part of his work--the Metaphysic of the Unconscious--our feet are clogged at every step. We are encircled by the merest play of words, the most unsatisfactory demonstrations, and most inconsistent inferences. The theory of final causes has been hitherto employed to show the wisdom of the world; with our Pessimist philosopher it shows nothing but its irrationality and misery. Consciousness has been generally supposed to be the condition of all happiness and interest in life; here it simply awakens us to misery, and the lower an animal lies in the scale of conscious life, the better and the pleasanter its lot.

. . . . .

"Thus, then, the universe, as an emanation of the unconscious, has been constructed. {90c} Throughout it has been marked by design, by purpose, by finality; throughout a wonderful adaptation of means to ends, a wonderful adjustment and relativity in different portions has been noticed--and all this for what conclusion? Not, as in the hands of the natural theologians of the eighteenth century, to show that the world is the result of design, of an intelligent, beneficent Creator, but the manifestation of a Being whose only predicates are negatives, whose very essence is to be unconscious. It is not only like ancient Athens, to an unknown, but to an unknowing God, that modern Pessimism rears its altar. Yet surely the fact that the motive principle of existence moves in a mysterious way outside our consciousness no way requires that the All-one Being should be himself unconscious.

I believe the foregoing to convey as correct an idea of Von Hartmann's system as it is possible to convey, and will leave it to the reader to say how much in common there is between this and the lecture given in the preceding chapter, beyond the fact that both touch upon unconscious actions. The extract which will form my next chapter is only about a thirtieth part of the entire "Philosophy of the Unconscious," but it will, I believe, suffice to substantiate the justice of what Mr. Sully has said in the passages above quoted.

As regards the accuracy of the translation, I have submitted all passages about which I was in the least doubtful to the same gentleman who revised my translation of Professor Hering's lecture; I have also given the German wherever I thought the reader might be glad to see it.

CHAPTER VIII

Translation of the chapter on "The Unconscious in Instinct," from Von Hartmann's "Philosophy of the Unconscious."

Von Hartmann's chapter on instinct is as follows:-

Instinct is action taken in pursuance of a purpose but without conscious perception of what the purpose is. {92a}

A purposive action, with consciousness of the purpose and where the course taken is the result of deliberation is not said to be instinctive; nor yet, again, is blind aimless action, such as outbreaks of fury on the part of offended or otherwise enraged animals. I see no occasion for disturbing the commonly received definition of instinct as given above; for those who think they can refer all the so-called ordinary instincts of animals to conscious deliberation ipso facto deny that there is such a thing as instinct at all, and should strike the word out of their vocabulary. But of this more hereafter.

Assuming, then, the existence of instinctive action as above defined, it can be explained as -

I. A mere necessary consequence of bodily organisation. {92b}

II. A mechanism of brain or mind contrived by nature.

III. The outcome of an unconscious activity of mind.

In neither of the two first cases is there any scope for the idea of purpose; in the third, purpose must be present immediately before the action. In the two first cases, action is supposed to be brought about by means of an initial arrangement, either of bodily or mental mechanism, purpose being conceived of as existing on a single occasion only--that is to say, in the determination of the initial arrangement. In the third, purpose is conceived as present in every individual instance. Let us proceed to the consideration of these three cases.

Instinct is not a mere consequence of bodily organisation; for -

(a.) Bodies may be alike, yet they may be endowed with different instincts.

All spiders have the same spinning apparatus, but one kind weaves radiating webs, another irregular ones, while a third makes none at all, but lives in holes, whose walls it overspins, and whose entrance it closes with a door. Almost all birds have a like organisation for the construction of their nests (a beak and feet), but how infinitely do their nests vary in appearance, mode of construction, attachment to surrounding objects (they stand, are glued on, hang, &c.), selection of site (caves, holes, corners, forks of trees, shrubs, the ground), and excellence of workmanship; how often, too, are they not varied in the species of a single genus, as of parus. Many birds, moreover, build no nest at all. The difference in the songs of birds are in like manner independent of the special construction of their voice apparatus, nor do the modes of nest construction that obtain among ants and bees depend upon their bodily organisation. Organisation, as a general rule, only renders the bird capable of singing, as giving it an apparatus with which to sing at all, but it has nothing to do with the specific character of the execution . . .

The nursing, defence, and education of offspring cannot be considered as in any way more dependent upon bodily organisation; nor yet the sites which insects choose for the laying of their eggs; nor, again, the selection of deposits of spawn, of their own species, by male fish for impregnation. The rabbit burrows, the hare does not, though both have the same burrowing apparatus. The hare, however, has less need of a subterranean place of refuge by reason of its greater swiftness. Some birds, with excellent powers of flight, are nevertheless stationary in their habits, as the secretary falcon and certain other birds of prey; while even such moderate fliers as quails are sometimes known to make very distant migrations.

(b.) Like instincts may be found associated with unlike organs.

Birds with and without feet adapted for climbing live in trees; so also do monkeys with and without flexible tails, squirrels, sloths, pumas, &c. Mole-crickets dig with a well-pronounced spade upon their fore-feet, while the burying-beetle does the same thing though it has no special apparatus whatever. The mole conveys its winter provender in pockets, an inch wide, long and half an inch wide within its cheeks; the field-mouse does so without the help of any such contrivance. The migratory instinct displays itself with equal strength in animals of widely different form, by whatever means they may pursue their journey, whether by water, land, or air.

It is clear, therefore, that instinct is in great measure independent of bodily organisation. Granted, indeed, that a certain amount of bodily apparatus is a sine qua non for any power of execution at all-
-as, for example, that there would be no ingenious nest without organs more or less adapted for its construction, no spinning of a web without spinning glands--nevertheless, it is impossible to maintain that instinct is a consequence of organisation. The mere existence of the organ does not constitute even the smallest

incentive to any corresponding habitual activity. A sensation of
pleasure must at least accompany the use of the organ before its
existence can incite to its employment. And even so when a sensation
of pleasure has given the impulse which is to render it active, it is
only the fact of there being activity at all, and not the special
characteristics of the activity, that can be due to organisation.
The reason for the special mode of the activity is the very problem
that we have to solve. No one will call the action of the spider
instinctive in voiding the fluid from her spinning gland when it is
too full, and therefore painful to her; nor that of the male fish
when it does what amounts to much the same thing as this. The
instinct and the marvel lie in the fact that the spider spins
threads, and proceeds to weave her web with them, and that the male
fish will only impregnate ova of his own species.

Another proof that the pleasure felt in the employment of an organ is
wholly inadequate to account for this employment is to be found in
the fact that the moral greatness of instinct, the point in respect
of which it most commands our admiration, consists in the obedience
paid to its behests, to the postponement of all personal well-being,
and at the cost, it may be, of life itself. If the mere pleasure of
relieving certain glands from overfulness were the reason why
caterpillars generally spin webs, they would go on spinning until
they had relieved these glands, but they would not repair their work
as often as any one destroyed it, and do this again and again until
they die of exhaustion. The same holds good with the other instincts
that at first sight appear to be inspired only by a sensation of
pleasure; for if we change the circumstances, so as to put self-
sacrifice in the place of self-interest, it becomes at once apparent
that they have a higher source than this. We think, for example,
that birds pair for the sake of mere sexual gratification; why, then,
do they leave off pairing as soon as they have laid the requisite
number of eggs? That there is a reproductive instinct over and above

the desire for sexual gratification appears from the fact that if a man takes an egg out of the nest, the birds will come together again and the hen will lay another egg; or, if they belong to some of the more wary species, they will desert their nest, and make preparation for an entirely new brood. A female wryneck, whose nest was daily robbed of the egg she laid in it, continued to lay a new one, which grew smaller and smaller, till, when she had laid her twenty-ninth egg, she was found dead upon her nest. If an instinct cannot stand the test of self-sacrifice--if it is the simple outcome of a desire for bodily gratification--then it is no true instinct, and is only so called erroneously.

Instinct is not a mechanism of brain or mind implanted in living beings by nature; for, if it were, then instinctive action without any, even unconscious, activity of mind, and with no conception concerning the purpose of the action, would be executed mechanically, the purpose having been once for all thought out by Nature or Providence, which has so organised the individual that it acts henceforth as a purely mechanical medium. We are now dealing with a psychical organisation as the cause instinct, as we were above dealing with a physical. psychical organisation would be a conceivable explanation and we need look no farther if every instinct once belonging to an animal discharged its functions in an unvarying manner. But this is never found to be the case, for instincts vary when there arises a sufficient motive for varying them. This proves that special exterior circumstances enter into the matter, and that these circumstances are the very things that render the attainment of the purpose possible through means selected by the instinct. Here first do we find instinct acting as though it were actually design with action following at its heels, for until the arrival of the motive, the instinct remains late and discharges no function whatever. The motive enters by way of an idea received into the mind through the instrumentality of the senses, and there is a constant connection between instinct in action and all sensual images which give information that an opportunity has arisen for attaining the ends proposed to itself by the instinct.

The psychical mechanism of this constant connection must also be looked for. It may help us here to turn to the piano for an illustration. The struck keys are the motives, the notes that sound in consequence are the instincts in action. This illustration might perhaps be allowed to pass (if we also suppose that entirely different keys can give out the same sound) if instincts could only be compared with DISTINCTLY TUNED notes, so that one and the same instinct acted always in the same manner on the rising of the motive which should set it in action. This, however, is not so; for it is the blind unconscious purpose of the instinct that is alone constant, the instinct itself--that is to say, the will to make use of certain means--varying as the means that can be most suitably employed vary under varying circumstances.

In this we condemn the theory which refuses to recognise unconscious purpose as present in each individual case of instinctive action. For he who maintains instinct to be the result of a mechanism of mind, must suppose a special and constant mechanism for each variation and modification of the instinct in accordance with exterior circumstances, {97} that is to say, a new string giving a note with a new tone must be inserted, and this would involve the mechanism in endless complication. But the fact that the purpose is constant notwithstanding all manner of variation in the means chosen by the instinct, proves that there is no necessity for the supposition of such an elaborate mental mechanism--the presence of an unconscious purpose being sufficient to explain the facts. The purpose of the bird, for example, that has laid her eggs is constant, and consists in the desire to bring her young to maturity. When the temperature of the air is insufficient to effect this, she sits upon her eggs, and only intermits her sittings in the warmest countries; the mammal, on the other hand, attains the fulfilment of its instinctive purpose without any co-operation on its own part. In warm climates many birds only sit by night, and small exotic birds that have built in aviaries kept at a high temperature sit little upon their eggs or not at all. How inconceivable is the supposition of a mechanism that impels the bird to sit as soon as the temperature falls below a certain height! How clear and simple, on the other hand, is the view that there is an unconscious purpose constraining the volition of the bird to the use of the fitting means, of which process, however, only the last link, that is to say, the will immediately preceding the action falls within the consciousness of the bird!

In South Africa the sparrow surrounds her nest with thorns as a defence against apes and serpents. The eggs of the cuckoo, as regards size, colour, and marking, invariably resemble those of the birds in whose nests she lays. Sylvia ruja, for example, lays a white egg with violet spots; Sylvia hippolais, a red one with black spots; Regulus ignicapellus, a cloudy red; but the cuckoo's egg is in each case so deceptive an imitation of its model, that it can hardly be distinguished except by the structure of its shell.

Huber contrived that his bees should be unable to build in their usual instinctive manner, beginning from above and working downwards; on this they began building from below, and again horizontally. The outermost cells that spring from the top of the hive or abut against its sides are not hexagonal, but pentagonal, so as to gain in strength, being attached with one base instead of two sides. In autumn bees lengthen their existing honey cells if these are insufficient, but in the ensuing spring they again shorten them in order to get greater roadway between the combs. When the full combs have become too heavy, they strengthen the walls of the uppermost or

bearing cells by thickening them with wax and propolis. If larvae of
working bees are introduced into the cells set apart for drones, the
working bees will cover these cells with the flat lids usual for this
kind of larvae, and not with the round ones that are proper for
drones. In autumn, as a general rule, bees kill their drones, but
they refrain from doing this when they have lost their queen, and
keep them to fertilise the young queen, who will be developed from
larvae that would otherwise have become working bees. Huber observed
that they defend the entrance of their hive against the inroads of
the sphinx moth by means of skilful constructions made of wax and
propolis. They only introduce propolis when they want it for the
execution of repairs, or for some other special purpose. Spiders and
caterpillars also display marvellous dexterity in the repair of their
webs if they have been damaged, and this requires powers perfectly
distinct from those requisite for the construction of a new one.

The above examples might be multiplied indefinitely, but they are
sufficient to establish the fact that instincts are not capacities
rolled, as it were, off a reel mechanically, according to an
invariable system, but that they adapt themselves most closely to the
circumstances of each case, and are capable of such great
modification and variation that at times they almost appear to cease
to be instinctive.

Many will, indeed, ascribe these modifications to conscious
deliberation on the part of the animals themselves, and it is
impossible to deny that in the case of the more intellectually gifted
animals there may be such a thing as a combination of instinctive
faculty and conscious reflection. I think, however, the examples
already cited are enough to show that often where the normal and the
abnormal action springs from the same source, without any
complication with conscious deliberation, they are either both
instinctive or both deliberative. {99} Or is that which prompts the

bee to build hexagonal prisms in the middle of her comb something of an actually distinct character from that which impels her to build pentagonal ones at the sides? Are there two separate kinds of thing, one of which induces birds under certain circumstances to sit upon their eggs, while another leads them under certain other circumstances to refrain from doing so? And does this hold good also with bees when they at one time kill their brethren without mercy and at another grant them their lives? Or with birds when they construct the kind of nest peculiar to their race, and, again, any special provision which they may think fit under certain circumstances to take? If it is once granted that the normal and the abnormal manifestations of instinct--and they are often incapable of being distinguished--spring from a single source, then the objection that the modification is due to conscious knowledge will be found to be a suicidal one later on, so far as it is directed against instinct generally. It may be sufficient here to point out, in anticipation of remarks that will be found in later chapters, that instinct and the power of organic development involve the same essential principle, though operating under different circumstances--the two melting into one another without any definite boundary between them. Here, then, we have conclusive proof that instinct does not depend upon organisation of body or brain, but that, more truly, the organisation is due to the nature and manner of the instinct.

On the other hand, we must now return to a closer consideration of the conception of a psychical mechanism. {100} And here we find that this mechanism, in spite of its explaining so much, is itself so obscure that we can hardly form any idea concerning it. The motive enters the mind by way of a conscious sensual impression; this is the first link of the process; the last link {101} appears as the conscious motive of an action. Both, however, are entirely unlike, and neither has anything to do with ordinary motivation, which consists exclusively in the desire that springs from a conception either of pleasure or dislike--the former prompting to the attainment of any object, the latter to its avoidance. In the case of instinct, pleasure is for the most part a concomitant phenomenon; but it is not so always, as we have already seen, inasmuch as the consummation and highest moral development of instinct displays itself in self-sacrifice.

The true problem, however, lies far deeper than this. For every conception of a pleasure proves that we have experienced this pleasure already. But it follows from this, that when the pleasure was first felt there must have been will present, in the gratification of which will the pleasure consisted; the question, therefore, arises, whence did the will come before the pleasure that would follow on its gratification was known, and before bodily pain, as, for example, of hunger, rendered relief imperative? Yet we may see that even though an animal has grown up apart from any others of its kind, it will yet none the less manifest the instinctive impulses of its race, though experience can have taught it nothing whatever concerning the pleasure that will ensue upon their gratification. As regards instinct, therefore, there must be a causal connection between the motivating sensual conception and the will to perform the instinctive action, and the pleasure of the subsequent gratification has nothing to do with the matter. We know by the experience of our own instincts that this causal connection does not lie within our consciousness; {102a} therefore, if it is to be a mechanism of any kind, it can only be either an unconscious mechanical induction and metamorphosis of the vibrations of the conceived motive into the vibrations of the conscious action in the brain, or an unconscious spiritual mechanism.

In the first case, it is surely strange that this process should go on unconsciously, though it is so powerful in its effects that the will resulting from it overpowers every other consideration, every other kind of will, and that vibrations of this kind, when set up in the brain, become always consciously perceived; nor is it easy to conceive in what way this metamorphosis can take place so that the constant purpose can be attained under varying circumstances by the resulting will in modes that vary with variation of the special features of each individual case.

But if we take the other alternative, and suppose an unconscious mental mechanism, we cannot legitimately conceive of the process going on in this as other than what prevails in all mental mechanism, namely, than as by way of idea and will. We are, therefore, compelled to imagine a causal connection between the consciously recognised motive and the will to do the instinctive action, through unconscious idea and will; nor do I know how this connection can be conceived as being brought about more simply than through a conceived and willed purpose. {102b} Arrived at this point, however, we have attained the logical mechanism peculiar to and inseparable from all mind, and find unconscious purpose to be an indispensable link in every instinctive action. With this, therefore, the conception of a mental mechanism, dead and predestined from without, has disappeared, and has become transformed into the spiritual life inseparable from logic, so that we have reached the sole remaining requirement for the conception of an actual instinct, which proves to be a conscious willing of the means towards an unconsciously willed purpose. This conception explains clearly and without violence all the problems which instinct presents to us; or more truly, all that was problematical about instinct disappears when its true nature has been thus declared. If this work were confined to the consideration of instinct alone, the conception of an unconscious activity of mind might excite opposition, inasmuch as it is one with which our educated public is not yet familiar; but in a work like the present, every chapter of which adduces fresh facts in support of the existence of such an activity and of its remarkable consequences, the novelty of the theory should be taken no farther into consideration.

Though I so confidently deny that instinct is the simple action of a mechanism which has been contrived once for all, I by no means exclude the supposition that in the constitution of the brain, the ganglia, and the whole body, in respect of morphological as well as molecular-physiological condition, certain predispositions can be established which direct the unconscious intermediaries more readily into one channel than into another. This predisposition is either the result of a habit which keeps continually cutting for itself a deeper and deeper channel, until in the end it leaves indelible traces whether in the individual or in the race, or it is expressly called into being by the unconscious formative principle in generation, so as to facilitate action in a given direction. This last will be the case more frequently in respect of exterior organisation--as, for example, with the weapons or working organs of animals--while to the former must be referred the molecular condition of brain and ganglia which bring about the perpetually recurring elements of an instinct such as the hexagonal shape of the cells of bees. We shall presently see that by individual character we mean the sum of the individual methods of reaction against all possible motives, and that this character depends essentially upon a constitution of mind and body acquired in some measure through habit by the individual, but for the most part inherited. But an instinct is also a mode of reaction against certain motives; here, too, then, we are dealing with character, though perhaps not so much with that of the individual as of the race; for by character in regard to instinct we do not intend the differences that distinguish individuals, but races from one another. If any one chooses to maintain that such a predisposition for certain kinds of activity on the part of brain and body constitutes a mechanism, this may in one sense be admitted; but as against this view it must be remarked -

1. That such deviations from the normal scheme of an instinct as cannot be referred to conscious deliberation are not provided for by any predisposition in this mechanism.

2. That heredity is only possible under the circumstances of a constant superintendence of the embryonic development by a purposive unconscious activity of growth. It must be admitted, however, that this is influenced in return by the predisposition existing in the germ.

3. That the impressing of the predisposition upon the individual from whom it is inherited can only be effected by long practice, consequently the instinct without auxiliary mechanism {105a} is the originating cause of the auxiliary mechanism.

4. That none of those instinctive actions that are performed rarely, or perhaps once only, in the lifetime of any individual--as, for example, those connected with the propagation and metamorphoses of the lower forms of life, and none of those instinctive omissions of action, neglect of which necessarily entails death--can be conceived as having become engrained into the character through habit; the ganglionic constitution, therefore, that predisposes the animal towards them must have been fashioned purposively.

5. That even the presence of an auxiliary mechanism {105b} does not compel the unconscious to a particular corresponding mode of instinctive action, but only predisposes it. This is shown by the possibility of departure from the normal type of action, so that the unconscious purpose is always stronger than the ganglionic constitution, and takes any opportunity of choosing from several similar possible courses the one that is handiest and most convenient to the constitution of the individual.

We now approach the question that I have reserved for our final one,--Is there, namely, actually such a thing as instinct, {105c} or are all so-called instinctive actions only the results of conscious deliberation?

In support of the second of these two views, it may be alleged that the more limited is the range of the conscious mental activity of any living being, the more fully developed in proportion to its entire mental power is its performance commonly found to be in respect of its own limited and special instinctive department. This holds as good with the lower animals as with men, and is explained by the fact that perfection of proficiency is only partly dependent upon natural capacity, but is in great measure due to practice and cultivation of the original faculty. A philologist, for example, is unskilled in questions of jurisprudence; a natural philosopher or mathematician, in philology; an abstract philosopher, in poetical criticism. Nor has this anything to do with the natural talents of the several persons, but follows as a consequence of their special training. The more special, therefore, is the direction in which the mental activity of any living being is exercised, the more will the whole developing and practising power of the mind be brought to bear upon this one branch, so that it is not surprising if the special power comes ultimately to bear an increased proportion to the total power of the individual, through the contraction of the range within which it is exercised.

Those, however, who apply this to the elucidation of instinct should not forget the words, "in proportion to the entire mental power of the animal in question," and should bear in mind that the entire mental power becomes less and less continually as we descend the scale of animal life, whereas proficiency in the performance of an instinctive action seems to be much of a muchness in all grades of the animal world. As, therefore, those performances which indisputably proceed from conscious deliberation decrease proportionately with decrease of mental power, while nothing of the kind is observable in the case of instinct--it follows that instinct must involve some other principle than that of conscious intelligence. We see, moreover, that actions which have their source in conscious intelligence are of one and the same kind, whether among the lower animals or with mankind--that is to say, that they are acquired by apprenticeship or instruction and perfected by practice; so that the saying, "Age brings wisdom," holds good with the brutes as much as with ourselves. Instinctive actions, on the contrary, have a special and distinct character, in that they are performed with no less proficiency by animals that have been reared in solitude than by those that have been instructed by their parents, the first essays of a hitherto unpractised animal being as successful as its later ones. There is a difference in principle here which cannot be mistaken. Again, we know by experience that the feebler and more limited an intelligence is, the more slowly do ideas act upon it, that is to say, the slower and more laborious is its conscious thought. So long as instinct does not come into play, this holds good both in the case of men of different powers of comprehension and with animals; but with instinct all is changed, for it is the speciality of instinct never to hesitate or loiter, but to take action instantly upon perceiving that the stimulating motive has made its appearance. This rapidity in arriving at a resolution is common to the instinctive actions both of the highest and the lowest animals, and indicates an essential difference between instinct and conscious deliberation.

Finally, as regards perfection of the power of execution, a glance will suffice to show the disproportion that exists between this and the grade of intellectual activity on which an animal may be standing. Take, for instance, the caterpillar of the emperor moth (Saturnia pavonia minor). It eats the leaves of the bush upon which it was born; at the utmost has just enough sense to get on to the lower sides of the leaves if it begins to rain, and from time to time changes its skin. This is its whole existence, which certainly does not lead us to expect a display of any, even the most limited, intellectual power. When, however, the time comes for the larva of this moth to become a chrysalis, it spins for itself a double cocoon, fortified with bristles that point outwards, so that it can be opened easily from within, though it is sufficiently impenetrable from without. If this contrivance were the result of conscious reflection, we should have to suppose some such reasoning process as the following to take place in the mind of the caterpillar:- "I am about to become a chrysalis, and, motionless as I must be, shall be exposed to many different kinds of attack. I must therefore weave myself a web. But when I am a moth I shall not be able, as some moths are, to find my way out of it by chemical or mechanical means; therefore I must leave a way open for myself. In order, however, that my enemies may not take advantage of this, I will close it with elastic bristles, which I can easily push asunder from within, but which, upon the principle of the arch, will resist all pressure from without." Surely this is asking rather too much from a poor caterpillar; yet the whole of the foregoing must be thought out if a correct result is to be arrived at.

This theoretical separation of instinct from conscious intelligence can be easily misrepresented by opponents of my theory, as though a separation in practice also would be necessitated in consequence. This is by no means my intention. On the contrary, I have already insisted at some length that both the two kinds of mental activity may co-exist in all manner of different proportions, so that there may be every degree of combination, from pure instinct to pure deliberation. We shall see, however, in a later chapter, that even in the highest and most abstract activity of human consciousness there are forces at work that are of the highest importance, and are essentially of the same kind as instinct.

On the other hand, the most marvellous displays of instinct are to be found not only in plants, but also in those lowest organisms of the simplest bodily form which are partly unicellular, and in respect of conscious intelligence stand far below the higher plants--to which, indeed, any kind of deliberative faculty is commonly denied. Even in the case of those minute microscopic organisms that baffle our attempts to classify them either as animals or vegetables, we are still compelled to admire an instinctive, purposive behaviour, which goes far beyond a mere reflex responsive to a stimulus from without; all doubt, therefore, concerning the actual existence of an instinct must be at an end, and the attempt to deduce it as a consequence of conscious deliberation be given up as hopeless. I will here adduce an instance as extraordinary as any we yet know of, showing, as it does, that many different purposes, which in the case of the higher animals require a complicated system of organs of motion, can be attained with incredibly simple means.

Arcella vulgaris is a minute morsel of protoplasm, which lives in a concave-convex, brown, finely reticulated shell, through a circular opening in the concave side of which it can project itself by throwing out pseudopodia. If we look through the microscope at a drop of water containing living arcellae, we may happen to see one of them lying on its back at the bottom of the drop, and making fruitless efforts for two or three minutes to lay hold of some fixed point by means of a pseudopodium. After this there will appear suddenly from two to five, but sometimes more, dark points in the protoplasm at a small distance from the circumference, and, as a rule, at regular distances from one another. These rapidly develop themselves into well-defined spherical air vesicles, and come presently to fill a considerable part of the hollow of the shell, thereby driving part of the protoplasm outside it. After from five

to twenty minutes, the specific gravity of the arcella is so much
lessened that it is lifted by the water with its pseudopodia, and
brought up against the upper surface of the water-drop, on which it
is able to travel. In from five to ten minutes the vesicles will now
disappear, the last small point vanishing with a jerk. If, however,
the creature has been accidentally turned over during its journey,
and reaches the top of the water-drop with its back uppermost, the
vesicles will continue growing only on one side, while they diminish
on the other; by this means the shell is brought first into an
oblique and then into a vertical position, until one of the
pseudopodia obtains a footing and the whole turns over. From the
moment the animal has obtained foothold, the bladders become
immediately smaller, and after they have disappeared the experiment
may be repeated at pleasure.

The positions of the protoplasm which the vesicles fashion change
continually; only the grainless protoplasm of the pseudopodia
develops no air. After long and fruitless efforts a manifest fatigue
sets in; the animal gives up the attempt for a time, and resumes it
after an interval of repose.

Engelmann, the discoverer of these phenomena, says (Pfluger's Archiv
fur Physologie, Bd. II.): "The changes in volume in all the vesicles
of the same animal are for the most part synchronous, effected in the
same manner, and of like size. There are, however, not a few
exceptions; it often happens that some of them increase or diminish
in volume much faster than others; sometimes one may increase while
another diminishes; all the changes, however, are throughout
unquestionably intentional. The object of the air-vesicles is to
bring the animal into such a position that it can take fast hold of
something with its pseudopodia. When this has been obtained, the air
disappears without our being able to discover any other reason for
its disappearance than the fact that it is no longer needed. . . .

If we bear these circumstances in mind, we can almost always tell whether an arcella will develop air-vesicles or no; and if it has already developed them, we can tell whether they will increase or diminish . . . The arcellae, in fact, in this power of altering their specific gravity possess a mechanism for raising themselves to the top of the water, or lowering themselves to the bottom at will. They use this not only in the abnormal circumstances of their being under microscopical observation, but at all times, as may be known by our being always able to find some specimens with air-bladders at the top of the water in which they live."

If what has been already advanced has failed to convince the reader of the hopelessness of attempting to explain instinct as a mode of conscious deliberation, he must admit that the following considerations are conclusive. It is most certain that deliberation and conscious reflection can only take account of such data as are consciously perceived; if, then, it can be shown that data absolutely indispensable for the arrival at a just conclusion cannot by any possibility have been known consciously, the result can no longer be held as having had its source in conscious deliberation. It is admitted that the only way in which consciousness can arrive at a knowledge of exterior facts is by way of an impression made upon the senses. We must, therefore, prove that a knowledge of the facts indispensable for arrival at a just conclusion could not have been thus acquired. This may be done as follows: {111} for, Firstly, the facts in question lie in the future, and the present gives no ground for conjecturing the time and manner of their subsequent development.

Secondly, they are manifestly debarred from the category of perceptions perceived through the senses, inasmuch as no information can be derived concerning them except through experience of similar occurrences in time past, and such experience is plainly out of the question.

It would not affect the argument if, as I think likely, it were to turn out, with the advance of our physiological knowledge, that all the examples of the first case that I am about to adduce reduce themselves to examples of the second, as must be admitted to have already happened in respect of many that I have adduced hitherto. For it is hardly more difficult to conceive of a priori knowledge, disconnected from any impression made upon the senses, than of knowledge which, it is true, does at the present day manifest itself upon the occasion of certain general perceptions, but which can only be supposed to be connected with these by means of such a chain of inferences and judiciously applied knowledge as cannot be believed to exist when we have regard to the capacity and organisation of the animal we may be considering.

An example of the first case is supplied by the larva of the stag-beetle in its endeavour to make itself a convenient hole in which to become a chrysalis. The female larva digs a hole exactly her own size, but the male makes one as long again as himself, so as to allow for the growth of his horns, which will be about the same length as his body. A knowledge of this circumstance is indispensable if the result achieved is to be considered as due to reflection, yet the actual present of the larva affords it no ground for conjecturing beforehand the condition in which it will presently find itself.

As regards the second case, ferrets and buzzards fall forthwith upon blind worms or other non-poisonous snakes, and devour them then and there. But they exhibit the greatest caution in laying hold of adders, even though they have never before seen one, and will endeavour first to bruise their heads, so as to avoid being bitten. As there is nothing in any other respect alarming in the adder, a conscious knowledge of the danger of its bite is indispensable, if the conduct above described is to be referred to conscious deliberation. But this could only have been acquired through experience, and the possibility of such experience may be controlled in the case of animals that have been kept in captivity from their youth up, so that the knowledge displayed can be ascertained to be independent of experience. On the other hand, both the above illustrations afford evidence of an unconscious perception of the facts, and prove the existence of a direct knowledge underivable from any sensual impression or from consciousness.

This has always been recognised, {113} and has been described under the words "presentiment" or "foreboding." These words, however, refer, on the one hand, only to an unknowable in the future, separated from us by space, and not to one that is actually present; on the other hand, they denote only the faint, dull, indefinite echo returned by consciousness to an invariably distinct state of unconscious knowledge. Hence the word "presentiment," which carries with it an idea of faintness and indistinctness, while, however, it may be easily seen that sentiment destitute of all, even unconscious, ideas can have no influence upon the result, for knowledge can only follow upon an idea. A presentiment that sounds in consonance with our consciousness can indeed, under certain circumstances, become tolerably definite, so that in the case of man it can be expressed in thought and language; but experience teaches us that even among ourselves this is not so when instincts special to the human race come into play; we see rather that the echo of our unconscious knowledge which finds its way into our consciousness is so weak that it manifests itself only in the accompanying feelings or frame of mind, and represents but an infinitely small fraction of the sum of our sensations. It is obvious that such a faintly sympathetic consciousness cannot form a sufficient foundation for a superstructure of conscious deliberation; on the other hand, conscious deliberation would be unnecessary, inasmuch as the process of thinking must have been already gone through unconsciously, for every faint presentiment that obtrudes itself upon our consciousness is in fact only the consequence of a distinct unconscious knowledge, and the knowledge with which it is concerned is almost always an idea of the purpose of some instinctive action, or of one most intimately connected therewith. Thus, in the case of the stag-beetle, the purpose consists in the leaving space for the growth of the horns; the means, in the digging the hole of a sufficient size; and the unconscious knowledge, in prescience concerning the future development of the horns.

Lastly, all instinctive actions give us an impression of absolute security and infallibility. With instinct the will is never hesitating or weak, as it is when inferences are being drawn consciously. We never find instinct making mistakes; we cannot, therefore, ascribe a result which is so invariably precise to such an obscure condition of mind as is implied when the word presentiment is used; on the contrary, this absolute certainty is so characteristic a feature of instinctive actions, that it constitutes almost the only well-marked point of distinction between these and actions that are done upon reflection. But from this it must again follow that some principle lies at the root of instinct other than that which underlies reflective action, and this can only be looked for in a determination of the will through a process that lies in the unconscious, {115a} to which this character of unhesitating infallibility will attach itself in all our future investigations.

Many will be surprised at my ascribing to instinct an unconscious knowledge, arising out of no sensual impression, and yet invariably accurate. This, however, is not a consequence of my theory concerning instinct; it is the foundation on which that theory is based, and is forced upon us by facts. I must therefore adduce examples. And to give a name to the unconscious knowledge, which is not acquired through impression made upon the senses, but which will be found to be in our possession, though attained without the instrumentality of means, {115b} I prefer the word "clairvoyance" {115c} to "presentiment," which, for reasons already given, will not serve me. This word, therefore, will be here employed throughout, as above defined.

Let us now consider examples of the instincts of self-preservation, subsistence, migration, and the continuation of the species. Most animals know their natural enemies prior to experience of any hostile designs upon themselves. A flight of young pigeons, even though they have no old birds with them, will become shy, and will separate from one another on the approach of a bird of prey. Horses and cattle that come from countries where there are no lions become unquiet and display alarm as soon as they are aware that a lion is approaching them in the night. Horses going along a bridle-path that used to leave the town at the back of the old dens of the carnivora in the Berlin Zoological Gardens were often terrified by the propinquity of enemies who were entirely unknown to them. Sticklebacks will swim composedly among a number of voracious pike, knowing, as they do, that the pike will not touch them. For if a pike once by mistake swallows a stickleback, the stickleback will stick in its throat by reason of the spine it carries upon its back, and the pike must starve to death without being able to transmit his painful experience to his descendants. In some countries there are people who by choice eat dog's flesh; dogs are invariably savage in the presence of these persons, as recognising in them enemies at whose hands they may one day come to harm. This is the more wonderful inasmuch as dog's fat applied externally (as when rubbed upon boots) attracts dogs by its smell. Grant saw a young chimpanzee throw itself into convulsions of terror at the sight of a large snake; and even among ourselves a Gretchen can often detect a Mephistopheles. An insect of the genius bombyx will seize another of the genus parnopaea, and kill it wherever it finds it, without making any subsequent use of the body;

but we know that the last-named insect lies in wait for the eggs of the first, and is therefore the natural enemy of its race. The phenomenon known to stockdrivers and shepherds as "das Biesen des Viehes" affords another example. For when a "dassel" or "bies" fly draws near the herd, the cattle become unmanageable and run about among one another as though they were mad, knowing, as they do, that the larvae from the eggs which the fly will lay upon them will presently pierce their hides and occasion them painful sores. These "dassel" flies--which have no sting--closely resemble another kind of gadfly which has a sting. Nevertheless, this last kind is little feared by cattle, while the first is so to an inordinate extent. The laying of the eggs upon the skin is at the time quite painless, and no ill consequences follow until long afterwards, so that we cannot suppose the cattle to draw a conscious inference concerning the connection that exists between the two. I have already spoken of the foresight shown by ferrets and buzzards in respect of adders; in like manner a young honey-buzzard, on being shown a wasp for the first time, immediately devoured it after having squeezed the sting from its body. No animal, whose instinct has not been vitiated by unnatural habits, will eat poisonous plants. Even when apes have contracted bad habits through their having been brought into contact with mankind, they can still be trusted to show us whether certain fruits found in their native forests are poisonous or no; for if poisonous fruits are offered them they will refuse them with loud cries. Every animal will choose for its sustenance exactly those animal or vegetable substances which agree best with its digestive organs, without having received any instruction on the matter, and without testing them beforehand. Even, indeed, though we assume that the power of distinguishing the different kinds of food is due to sight and not to smell, it remains none the less mysterious how the animal can know what it is that will agree with it. Thus the kid which Galen took prematurely from its mother smelt at all the different kinds of food that were set before it, but drank only the

milk without touching anything else. The cherry-finch opens a
cherry-stone by turning it so that her beak can hit the part where
the two sides join, and does this as much with the first stone she
cracks as with the last. Fitchets, martens, and weasels make small
holes on the opposite sides of an egg which they are about to suck,
so that the air may come in while they are sucking. Not only do
animals know the food that will suit them best, but they find out the
most suitable remedies when they are ill, and constantly form a
correct diagnosis of their malady with a therapeutical knowledge
which they cannot possibly have acquired. Dogs will often eat a
great quantity of grass--particularly couch-grass--when they are
unwell, especially after spring, if they have worms, which thus pass
from them entangled in the grass, or if they want to get fragments of
bone from out of their stomachs. As a purgative they make use of
plants that sting. Hens and pigeons pick lime from walls and
pavements if their food does not afford them lime enough to make
their eggshells with. Little children eat chalk when suffering from
acidity of the stomach, and pieces of charcoal if they are troubled
with flatulence. We may observe these same instincts for certain
kinds of food or drugs even among grown-up people, under
circumstances in which their unconscious nature has unusual power;
as, for example, among women when they are pregnant, whose capricious
appetites are probably due to some special condition of the foetus,
which renders a certain state of the blood desirable. Field-mice
bite off the germs of the corn which they collect together, in order
to prevent its growing during the winter. Some days before the
beginning of cold weather the squirrel is most assiduous in
augmenting its store, and then closes its dwelling. Birds of passage
betake themselves to warmer countries at times when there is still no
scarcity of food for them here, and when the temperature is
considerably warmer than it will be when they return to us. The same
holds good of the time when animals begin to prepare their winter
quarters, which beetles constantly do during the very hottest days of

autumn. When swallows and storks find their way back to their native
places over distances of hundreds of miles, and though the aspect of
the country is reversed, we say that this is due to the acuteness of
their perception of locality; but the same cannot be said of dogs,
which, though they have been carried in a bag from one place to
another that they do not know, and have been turned round and round
twenty times over, have still been known to find their way home.
Here we can say no more than that their instinct has conducted them--
that the clairvoyance of the unconscious has allowed them to
conjecture their way. {119a}

Before an early winter, birds of passage collect themselves in
preparation for their flight sooner than usual; but when the winter
is going to be mild, they will either not migrate at all, or travel
only a small distance southward. When a hard winter is coming,
tortoises will make their burrows deeper. If wild geese, cranes,
etc., soon return from the countries to which they had betaken
themselves at the beginning of spring, it is a sign that a hot and
dry summer is about to ensue in those countries, and that the drought
will prevent their being able to rear their young. In years of
flood, beavers construct their dwellings at a higher level than
usual, and shortly before an inundation the field-mice in Kamtschatka
come out of their holes in large bands. If the summer is going to be
dry, spiders may be seen in May and April, hanging from the ends of
threads several feet in length. If in winter spiders are seen
running about much, fighting with one another and preparing new webs,
there will be cold weather within the next nine days, or from that to
twelve: when they again hide themselves there will be a thaw. I
have no doubt that much of this power of prophesying the weather is
due to a perception of certain atmospheric conditions which escape
ourselves, but this perception can only have relation to a certain
actual and now present condition of the weather; and what can the
impression made by this have to do with their idea of the weather

that will ensue? No one will ascribe to animals a power of
prognosticating the weather months beforehand by means of inferences
drawn logically from a series of observations, {119b} to the extent
of being able to foretell floods. It is far more probable that the
power of perceiving subtle differences of actual atmospheric
condition is nothing more than the sensual perception which acts as
motive--for a motive must assuredly be always present--when an
instinct comes into operation. It continues to hold good, therefore,
that the power of foreseeing the weather is a case of unconscious
clairvoyance, of which the stork which takes its departure for the
south four weeks earlier than usual knows no more than does the stag
when before a cold winter he grows himself a thicker pelt than is his
wont. On the one hand, animals have present in their consciousness a
perception of the actual state of the weather; on the other, their
ensuing action is precisely such as it would be if the idea present
with them was that of the weather that is about to come. This they
cannot consciously have; the only natural intermediate link,
therefore, between their conscious knowledge and their action is
supplied by unconscious idea, which, however, is always accurately
prescient, inasmuch as it contains something which is neither given
directly to the animal through sensual perception, nor can be deduced
inferentially through the understanding.

Most wonderful of all are the instincts connected with the
continuation of the species. The males always find out the females
of their own kind, but certainly not solely through their resemblance
to themselves. With many animals, as, for example, parasitic crabs,
the sexes so little resemble one another that the male would be more
likely to seek a mate from the females of a thousand other species
than from his own. Certain butterflies are polymorphic, and not only
do the males and females of the same species differ, but the females
present two distinct forms, one of which as a general rule mimics the
outward appearance of a distant but highly valued species; yet the

males will pair only with the females of their own kind, and not with the strangers, though these may be very likely much more like the males themselves. Among the insect species of the strepsiptera, the female is a shapeless worm which lives its whole life long in the hind body of a wasp; its head, which is of the shape of a lentil, protrudes between two of the belly rings of the wasp, the rest of the body being inside. The male, which only lives for a few hours, and resembles a moth, nevertheless recognises his mate in spite of these adverse circumstances, and fecundates her.

Before any experience of parturition, the knowledge that it is approaching drives all mammals into solitude, and bids them prepare a nest for their young in a hole or in some other place of shelter. The bird builds her nest as soon as she feels the eggs coming to maturity within her. Snails, land-crabs, tree-frogs, and toads, all of them ordinarily dwellers upon land, now betake themselves to the water; sea-tortoises go on shore, and many saltwater fishes come up into the rivers in order to lay their eggs where they can alone find the requisites for their development. Insects lay their eggs in the most varied kinds of situations,--in sand, on leaves, under the hides and horny substances of other animals; they often select the spot where the larva will be able most readily to find its future sustenance, as in autumn upon the trees that will open first in the coming spring, or in spring upon the blossoms that will first bear fruit in autumn, or in the insides of those caterpillars which will soonest as chrysalides provide the parasitic larva at once with food and with protection. Other insects select the sites from which they will first get forwarded to the destination best adapted for their development. Thus some horseflies lay their eggs upon the lips of horses or upon parts where they are accustomed to lick themselves. The eggs get conveyed hence into the entrails, the proper place for their development,--and are excreted upon their arrival at maturity. The flies that infest cattle know so well how to select the most vigorous and healthiest beasts, that cattle-dealers and tanners place entire dependence upon them, and prefer those beasts and hides that are most scarred by maggots. This selection of the best cattle by the help of these flies is no evidence in support of the conclusion that the flies possess the power of making experiments consciously and of reflecting thereupon, even though the men whose trade it is to do this recognise them as their masters. The solitary wasp makes a hole several inches deep in the sand, lays her egg, and packs along with it a number of green maggots that have no legs, and which, being on the point of becoming chrysalides, are well nourished and able to go a long time without food; she packs these maggots so closely together that they cannot move nor turn into chrysalides, and just enough of them to support the larva until it becomes a chrysalis. A kind of bug (cerceris bupresticida), which itself lives only upon pollen, lays her eggs in an underground cell, and with each one of them she deposits three beetles, which she has lain in wait for and captured when they were still weak through having only just left off being chrysalides. She kills these beetles, and appears to smear them with a fluid whereby she preserves them fresh and suitable for food. Many kinds of wasps open the cells in which their larvae are confined when these must have consumed the provision that was left with them. They supply them with more food, and again close the cell. Ants, again, hit always upon exactly the right moment for opening the cocoons in which their larvae are confined and for setting them free, the larva being unable to do this for itself. Yet the life of only a few kinds of insects lasts longer than a single breeding season. What then can they know about the contents of their eggs and the fittest place for their development? What can they know about the kind of food the larva will want when it leaves the egg--a food so different from their own? What, again, can they know about the quantity of food that will be necessary? How much of all this at least can they know consciously? Yet their actions, the pains they take, and the importance they evidently attach to these matters, prove that they have a foreknowledge of the future: this knowledge therefore can only be an unconscious clairvoyance. For clairvoyance it must certainly be that inspires the will of an animal to open cells and cocoons at the very moment that the larva is either ready for more food or fit for leaving the cocoon. The eggs of the cuckoo do not take only from two to three days to mature in her ovaries, as those of most birds do, but require from eleven to twelve; the cuckoo, therefore, cannot sit upon her own eggs, for her first egg would be spoiled before the last was laid. She therefore lays in other birds' nests--of course laying each egg in a different nest. But in order that the birds may not perceive her egg to be a stranger and turn it out of the nest, not only does she lay an egg much smaller than might be expected from a bird of her size (for she only finds her opportunity among small birds), but, as already said, she imitates the other eggs in the nest she has selected with surprising accuracy in respect both of colour and marking. As the cuckoo chooses the nest some days beforehand, it may be thought, if the nest is an open one, that the cuckoo looks upon the colour of the eggs within it while her own is in process of maturing inside her, and that it is thus her egg comes to assume the colour of the others; but this explanation will not hold good for nests that are made in the holes of trees, as that of sylvia phaenicurus, or which are oven-shaped with a narrow entrance, as with sylvia rufa. In these cases the cuckoo can neither slip in nor look in, and must therefore lay her egg outside the nest and push it inside with her beak; she can therefore have no means of perceiving through her senses what the eggs already in the nest are like. If, then, in spite of all this, her egg closely resembles the others, this can only have come about through an unconscious clairvoyance which directs the process that goes on within the ovary in respect of colour and marking.

An important argument in support of the existence of a clairvoyance in the instincts of animals is to be found in the series of facts which testify to the existence of a like clairvoyance, under certain circumstances, even among human beings, while the self-curative instincts of children and of pregnant women have been already mentioned. Here, however, {124} in correspondence with the higher stage of development which human consciousness has attained, a stronger echo of the unconscious clairvoyance commonly resounds within consciousness itself, and this is represented by a more or less definite presentiment of the consequences that will ensue. It is also in accord with the greater independence of the human intellect that this kind of presentiment is not felt exclusively immediately before the carrying out of an action, but is occasionally disconnected from the condition that an action has to be performed immediately, and displays itself simply as an idea independently of conscious will, provided only that the matter concerning which the presentiment is felt is one which in a high degree concerns the will of the person who feels it. In the intervals of an intermittent fever or of other illness, it not unfrequently happens that sick persons can accurately foretell the day of an approaching attack and how long it will last. The same thing occurs almost invariably in the case of spontaneous, and generally in that of artificial, somnambulism; certainly the Pythia, as is well known, used to announce the date of her next ecstatic state. In like manner the curative instinct displays itself in somnambulists, and they have been known to select remedies that have been no less remarkable for the success attending their employment than for the completeness with which they have run counter to received professional opinion. The indication of medicinal remedies is the only use which respectable electro-biologists will make of the half-sleeping, half-waking condition of those whom they are influencing. "People in perfectly sound health have been known, before childbirth or at the commencement of an illness, to predict accurately their own approaching death. The accomplishment of their predictions can hardly be explained as the result of mere chance, for if this were all, the prophecy should fail at least as often as not, whereas the reverse is actually the case. Many of these persons neither desire death nor fear it, so that the result cannot be ascribed to imagination." So writes the celebrated physiologist, Burdach, from whose chapter on presentiment in his work "Bhicke in's Leben" a great part of my most striking examples is taken. This presentiment of deaths, which is the exception among men, is quite common with animals, even though they do not know nor understand what death is. When they become aware that their end is approaching, they steal away to outlying and solitary places. This is why in cities we so rarely see the dead body or skeleton of a cat. We can only suppose that the unconscious clairvoyance, which is of essentially the same kind whether in man or beast, calls forth presentiments of different degrees of definiteness, so that the cat is driven to withdraw herself through a mere instinct without knowing why she does so, while in man a definite perception is awakened of the fact that he is about to die. Not only do people have presentiments concerning their own death, but there are many instances on record in which they have become aware of that of those near and dear to them, the dying person having appeared in a dream to friend or wife or husband. Stories to this effect prevail among all nations, and unquestionably contain much truth. Closely connected with this is the power of second sight, which existed formerly in Scotland, and still does so in the Danish islands. This power enables certain people without any ecstasy, but simply through their keener perception, to foresee coming events, or to tell what is going on in foreign countries on matters in which they are deeply interested, such as deaths, battles, conflagrations (Swedenborg foretold the burning of Stockholm), the arrival or the doings of friends who are at a distance. With many persons this clairvoyance is confined to a knowledge of the death of their acquaintances or fellow-townspeople. There have been a great many instances of such death-prophetesses, and, what is most important, some cases have been verified in courts of law. I may

say, in passing, that this power of second sight is found in persons
who are in ecstatic states, in the spontaneous or artificially
induced somnambulism of the higher kinds of waking dreams, as well as
in lucid moments before death. These prophetic glimpses, by which
the clairvoyance of the unconscious reveals itself to consciousness,
{126} are commonly obscure because in the brain they must assume a
form perceptible by the senses, whereas the unconscious idea can have
nothing to do with any form of sensual impression: it is for this
reason that humours, dreams, and the hallucinations of sick persons
can so easily have a false signification attached to them. The
chances of error and self-deception that arise from this source, the
ease with which people may be deceived intentionally, and the
mischief which, as a general rule, attends a knowledge of the future,
these considerations place beyond all doubt the practical unwisdom of
attempts to arrive at certainty concerning the future. This,
however, cannot affect the weight which in theory should be attached
to phenomena of this kind, and must not prevent us from recognising
the positive existence of the clairvoyance whose existence I am
maintaining, though it is often hidden under a chaos of madness and
imposture.

The materialistic and rationalistic tendencies of the present day
lead most people either to deny facts of this kind in toto, or to
ignore them, inasmuch as they are inexplicable from a materialistic
standpoint, and cannot be established by the inductive or
experimental method--as though this last were not equally impossible
in the case of morals, social science, and politics. A mind of any
candour will only be able to deny the truths of this entire class of
phenomena so long as it remains in ignorance of the facts that have
been related concerning them; but, again, a continuance in this
ignorance can only arise from unwillingness to be convinced. I am
satisfied that many of those who deny all human power of divination
would come to another, and, to say the least, more cautious

conclusion if they would be at the pains of further investigation; and I hold that no one, even at the present day, need be ashamed of joining in with an opinion which was maintained by all the great spirits of antiquity except Epicurus--an opinion whose possible truth hardly one of our best modern philosophers has ventured to contravene, and which the champions of German enlightenment were so little disposed to relegate to the domain of old wives' tales, that Goethe furnishes us with an example of second sight that fell within his own experience, and confirms it down to its minutest details.

Although I am far from believing that the kind of phenomena above referred to form in themselves a proper foundation for a superstructure of scientific demonstration, I nevertheless find them valuable as a completion and further confirmation of the series of phenomena presented to us by the clairvoyance which we observe in human and animal instinct. Even though they only continue this series {128} through the echo that is awakened within our consciousness, they as powerfully support the account which instinctive actions give concerning their own nature, as they are themselves supported by the analogy they present to the clairvoyance observable in instinct. This, then, as well as my desire not to lose an opportunity of protesting against a modern prejudice, must stand as my reason for having allowed myself to refer, in a scientific work, to a class of phenomena which has fallen at present into so much discredit.

I will conclude with a few words upon a special kind of instinct which has a very instructive bearing upon the subject generally, and shows how impossible it is to evade the supposition of an unconscious clairvoyance on the part of instinct. In the examples adduced hitherto, the action of each individual has been done on the individual's own behalf, except in the case of instincts connected with the continuation of the species, where the action benefits others--that is to say, the offspring of the creature performing it.

We must now examine the cases in which a solidarity of instinct is found to exist between several individuals, so that, on the one hand, the action of each redounds to the common welfare, and, on the other, it becomes possible for a useful purpose to be achieved through the harmonious association of individual workers. This community of instinct exists also among the higher animals, but here it is harder to distinguish from associations originating through conscious will, inasmuch as speech supplies the means of a more perfect intercommunication of aim and plan. We shall, however, definitely recognise {129} this general effect of a universal instinct in the origin of speech and in the great political and social movements in the history of the world. Here we are concerned only with the simplest and most definite examples that can be found anywhere, and therefore we will deal in preference with the lower animals, among which, in the absence of voice, the means of communicating thought, mimicry, and physiognomy, are so imperfect that the harmony and interconnection of the individual actions cannot in its main points be ascribed to an understanding arrived at through speech. Huber observed that when a new comb was being constructed a number of the largest working-bees, that were full of honey, took no part in the ordinary business of the others, but remained perfectly aloof. Twenty-four hours afterwards small plates of wax had formed under their bellies. The bee drew these off with her hind-feet, masticated them, and made them into a band. The small plates of wax thus prepared were then glued to the roof of the hive one on the top of the other. When one of the bees of this kind had used up her plates of wax, another followed her and carried the same work forward in the same way. A thin rough vertical wall, half a line in thickness and fastened to the sides of the hive, was thus constructed. On this, one of the smaller working-bees whose belly was empty came, and after surveying the wall, made a flat half-oval excavation in the middle of one of its sides; she piled up the wax thus excavated round the edge of the excavation. After a short time she was relieved by another like herself, till more than twenty followed one another in this way. Meanwhile another bee began to make a similar hollow on the other side of the wall, but corresponding only with the rim of the excavation on this side. Presently another bee began a second hollow upon the same side, each bee being continually relieved by others. Other bees kept coming up and bringing under their bellies plates of wax, with which they heightened the edge of the small wall of wax. In this, new bees were constantly excavating the ground for more cells, while others proceeded by degrees to bring those already begun into a perfectly symmetrical shape, and at the same time continued building up the prismatic walls between them. Thus the bees worked on opposite sides of the wall of wax, always on the same plan and in the closest correspondence with those upon the other side, until eventually the cells on both sides were completed in all their wonderful regularity and harmony of arrangement, not merely as regards those standing side by side, but also as regards those which were upon the other side of their pyramidal base.

Let the reader consider how animals that are accustomed to confer together, by speech or otherwise, concerning designs which they may be pursuing in common, will wrangle with thousandfold diversity of opinion; let him reflect how often something has to be undone, destroyed, and done over again; how at one time too many hands come forward, and at another too few; what running to and fro there is before each has found his right place; how often too many, and again too few, present themselves for a relief gang; and how we find all this in the concerted works of men, who stand so far higher than bees in the scale of organisation. We see nothing of the kind among bees. A survey of their operations leaves rather the impression upon us as though an invisible master-builder had prearranged a scheme of action for the entire community, and had impressed it upon each individual

member, as though each class of workers had learnt their appointed
work by heart, knew their places and the numbers in which they should
relieve each other, and were informed instantaneously by a secret
signal of the moment when their action was wanted. This, however, is
exactly the manner in which an instinct works; and as the intention
of the entire community is instinctively present in the unconscious
clairvoyance {131a} of each individual bee, so the possession of this
common instinct impels each one of them to the discharge of her
special duties when the right moment has arrived. It is only thus
that the wonderful tranquillity and order which we observe could be
attained. What we are to think concerning this common instinct must
be reserved for explanation later on, but the possibility of its
existence is already evident, inasmuch {131b} as each individual has
an unconscious insight concerning the plan proposed to itself by the
community, and also concerning the means immediately to be adopted
through concerted action--of which, however, only the part requiring
his own co-operation is present in the consciousness of each. Thus,
for example, the larva of the bee itself spins the silky chamber in
which it is to become a chrysalis, but other bees must close it with
its lid of wax. The purpose of there being a chamber in which the
larva can become a chrysalis must be present in the minds of each of
these two parties to the transaction, but neither of them acts under
the influence of conscious will, except in regard to his own
particular department. I have already mentioned the fact that the
larva, after its metamorphosis, must be freed from its cell by other
bees, and have told how the working-bees in autumn kill the drones,
so that they may not have to feed a number of useless mouths
throughout the winter, and how they only spare them when they are
wanted in order to fecundate a new queen. Furthermore, the working-
bees build cells in which the eggs laid by the queen may come to
maturity, and, as a general rule, make just as many chambers as the
queen lays eggs; they make these, moreover, in the same order as that
in which the queen lays her eggs, namely, first for the working-bees,

then for the drones, and lastly for the queens. In the polity of the
bees, the working and the sexual capacities, which were once united,
are now personified in three distinct kinds of individual, and these
combine with an inner, unconscious, spiritual union, so as to form a
single body politic, as the organs of a living body combine to form
the body itself.

In this chapter, therefore, we have arrived at the following
conclusions:-

Instinct is not the result of conscious deliberation; {132} it is not
a consequence of bodily organisation; it is not a mere result of a
mechanism which lies in the organisation of the brain; it is not the
operation of dead mechanism, glued on, as it were, to the soul, and
foreign to its inmost essence; but it is the spontaneous action of
the individual, springing from his most essential nature and
character. The purpose to which any particular kind of instinctive
action is subservient is not the purpose of a soul standing outside
the individual and near akin to Providence--a purpose once for all
thought out, and now become a matter of necessity to the individual,
so that he can act in no other way, though it is engrafted into his
nature from without, and not natural to it. The purpose of the
instinct is in each individual case thought out and willed
unconsciously by the individual, and afterwards the choice of means
adapted to each particular case is arrived at unconsciously. A
knowledge of the purpose is often absolutely unattainable {133} by
conscious knowledge through sensual perception. Then does the
peculiarity of the unconscious display itself in the clairvoyance of
which consciousness perceives partly only a faint and dull, and
partly, as in the case of man, a more or less definite echo by way of
sentiment, whereas the instinctive action itself--the carrying out of
the means necessary for the achievement of the unconscious purpose--
falls always more clearly within consciousness, inasmuch as due

performance of what is necessary would be otherwise impossible. Finally, the clairvoyance makes itself perceived in the concerted action of several individuals combining to carry out a common but unconscious purpose.

Up to this point we have encountered clairvoyance as a fact which we observe but cannot explain, and the reader may say that he prefers to take his stand here, and be content with regarding instinct simply as a matter of fact, the explanation of which is at present beyond our reach. Against this it must be urged, firstly, that clairvoyance is not confined to instinct, but is found also in man; secondly, that clairvoyance is by no means present in all instincts, and that therefore our experience shows us clairvoyance and instinct as two distinct things--clairvoyance being of great use in explaining instinct, but instinct serving nothing to explain clairvoyance; thirdly and lastly, that the clairvoyance of the individual will not continue to be so incomprehensible to us, but will be perfectly well explained in the further course of our investigation, while we must give up all hope of explaining instinct in any other way.

The conception we have thus arrived at enables us to regard instinct as the innermost kernel, so to speak, of every living being. That this is actually the case is shown by the instincts of self-preservation and of the continuation of the species which we observe throughout creation, and by the heroic self-abandonment with which the individual will sacrifice welfare, and even life, at the bidding of instinct. We see this when we think of the caterpillar, and how she repairs her cocoon until she yields to exhaustion; of the bird, and how she will lay herself to death; of the disquiet and grief displayed by all migratory animals if they are prevented from migrating. A captive cuckoo will always die at the approach of winter through despair at being unable to fly away; so will the vineyard snail if it is hindered of its winter sleep. The weakest mother will encounter an enemy far surpassing her in strength, and suffer death cheerfully for her offspring's sake. Every year we see fresh cases of people who have been unfortunate going mad or committing suicide. Women who have survived the Caesarian operation allow themselves so little to be deterred from further childbearing through fear of this frightful and generally fatal operation, that they will undergo it no less than three times. Can we suppose that what so closely resembles demoniacal possession can have come about through something engrafted on to the soul as a mechanism foreign to its inner nature, {135} or through conscious deliberation which adheres always to a bare egoism, and is utterly incapable of such self-sacrifice for the sake of offspring as is displayed by the procreative and maternal instincts?

We have now, finally, to consider how it arises that the instincts of any animal species are so similar within the limits of that species-- a circumstance which has not a little contributed to the engrafted-mechanism theory. But it is plain that like causes will be followed by like effects; and this should afford sufficient explanation. The bodily mechanism, for example, of all the individuals of a species is alike; so again are their capabilities and the outcomes of their conscious intelligence--though this, indeed, is not the case with man, nor in some measure even with the highest animals; and it is through this want of uniformity that there is such a thing as individuality. The external conditions of all the individuals of a species are also tolerably similar, and when they differ essentially, the instincts are likewise different--a fact in support of which no examples are necessary. From like conditions of mind and body (and this includes like predispositions of brain and ganglia) and like exterior circumstances, like desires will follow as a necessary logical consequence. Again, from like desires and like inward and outward circumstances, a like choice of means--that is to say, like instincts--must ensue. These last two steps would not be conceded without restriction if the question were one involving conscious deliberation, but as these logical consequences are supposed to follow from the unconscious, which takes the right step unfailingly without vacillation or delay so long as the premises are similar, the ensuing desires and the instincts to adopt the means for their gratification will be similar also.

Thus the view which we have taken concerning instinct explains the very last point which it may be thought worth while to bring forward in support of the opinions of our opponents.

I will conclude this chapter with the words of Schelling: "Thoughtful minds will hold the phenomena of animal instinct to belong to the most important of all phenomena, and to be the true touchstone of a durable philosophy."

CHAPTER IX

Remarks upon Von Hartmann's position in regard to instinct.

Uncertain how far the foregoing chapter is not better left without comment of any kind, I nevertheless think that some of my readers may be helped by the following extracts from the notes I took while translating. I will give them as they come, without throwing them into connected form.

Von Hartmann defines instinct as action done with a purpose, but without consciousness of purpose.

The building of her nest by a bird is an instinctive action; it is
done with a purpose, but it is arbitrary to say that the bird has no
knowledge of that purpose. Some hold that birds when they are
building their nest know as well that they mean to bring up a family
in it as a young married couple do when they build themselves a
house. This is the conclusion which would be come to by a plain
person on a prima facie view of the facts, and Von Hartmann shows no
reason for modifying it.

A better definition of instinct would be that it is inherited
knowledge in respect of certain facts, and of the most suitable
manner in which to deal with them.

Von Hartmann speaks of "a mechanism of brain or mind" contrived by
nature, and again of "a psychical organisation," as though it were
something distinct from a physical organisation.

We can conceive of such a thing as mechanism of brain, for we have
seen brain and handled it; but until we have seen a mind and handled
it, or at any rate been enabled to draw inferences which will warrant
us in conceiving of it as a material substance apart from bodily
substance, we cannot infer that it has an organisation apart from
bodily organisation. Does Von Hartmann mean that we have two bodies-
-a body-body, and a soul-body?

He says that no one will call the action of the spider instinctive in
voiding the fluids from its glands when they are too full. Why not?

He is continually personifying instinct; thus he speaks of the "ends proposed to itself by the instinct," of "the blind unconscious purpose of the instinct," of "an unconscious purpose constraining the volition of the bird," of "each variation and modification of the instinct," as though instinct, purpose, and, later on, clairvoyance, were persons, and not words characterising a certain class of actions. The ends are proposed to itself by the animal, not by the instinct. Nothing but mischief can come of a mode of expression which does not keep this clearly in view.

It must not be supposed that the same cuckoo is in the habit of laying in the nests of several different species, and of changing the colour of her eggs according to that of the eggs of the bird in whose nest she lays. I have inquired from Mr. R. Bowdler Sharpe of the ornithological department at the British Museum, who kindly gives it me as his opinion that though cuckoos do imitate the eggs of the species on whom they foist their young ones, yet one cuckoo will probably lay in the nests of one species also, and will stick to that species for life. If so, the same race of cuckoos may impose upon the same species for generations together. The instinct will even thus remain a very wonderful one, but it is not at all inconsistent with the theory put forward by Professor Hering and myself.

Returning to the idea of psychical mechanism, he admits that "it is itself so obscure that we can hardly form any idea concerning it," {139a} and then goes on to claim for it that it explains a great many other things. This must have been the passage which Mr. Sully had in view when he very justly wrote that Von Hartmann "dogmatically closes the field of physical inquiry, and takes refuge in a phantom which explains everything, simply because it is itself incapable of explanation."

According to Von Hartmann {139b} the unpractised animal manifests its
instinct as perfectly as the practised. This is not the case. The
young animal exhibits marvellous proficiency, but it gains by
experience. I have watched sparrows, which I can hardly doubt to be
young ones, spend a whole month in trying to build their nest, and
give it up in the end as hopeless. I have watched three such cases
this spring in a tree not twenty feet from my own window and on a
level with my eye, so that I have been able to see what was going on
at all hours of the day. In each case the nest was made well and
rapidly up to a certain point, and then got top-heavy and tumbled
over, so that little was left on the tree: it was reconstructed and
reconstructed over and over again, always with the same result, till
at last in all three cases the birds gave up in despair. I believe
the older and stronger birds secure the fixed and best sites, driving
the younger birds to the trees, and that the art of building nests in
trees is dying out among house-sparrows.

He declares that instinct is not due to organisation so much as
organisation to instinct. {140} The fact is, that neither can claim
precedence of or pre-eminence over the other. Instinct and
organisation are only mind and body, or mind and matter; and these
are not two separable things, but one and inseparable, with, as it
were, two sides; the one of which is a function of the other. There
was never yet either matter without mind, however low, nor mind,
however high, without a material body of some sort; there can be no
change in one without a corresponding change in the other; neither
came before the other; neither can either cease to change or cease to
be; for "to be" is to continue changing, so that "to be" and "to
change" are one.

Whence, he asks, comes the desire to gratify an instinct before experience of the pleasure that will ensue on gratification? This is a pertinent question, but it is met by Professor Hering with the answer that this is due to memory--to the continuation in the germ of vibrations that were vibrating in the body of the parent, and which, when stimulated by vibrations of a suitable rhythm, become more and more powerful till they suffice to set the body in visible action. For my own part I only venture to maintain that it is due to memory, that is to say, to an enduring sense on the part of the germ of the action it took when in the persons of its ancestors, and of the gratification which ensued thereon. This meets Von Hartmann's whole difficulty.

The glacier is not snow. It is snow packed tight into a small compass, and has thus lost all trace of its original form. How incomplete, however, would be any theory of glacial action which left out of sight the origin of the glacier in snow! Von Hartmann loses sight of the origin of instinctive in deliberative actions because the two classes of action are now in many respects different. His philosophy of the unconscious fails to consider what is the normal process by means of which such common actions as we can watch, and whose history we can follow, have come to be done unconsciously.

He says, {141} "How inconceivable is the supposition of a mechanism, &c., &c.; how clear and simple, on the other hand, is the view that there is an unconscious purpose constraining the volition of the bird to the use of the fitting means." Does he mean that there is an actual thing--an unconscious purpose--something outside the bird, as it were a man, which lays hold of the bird and makes it do this or that, as a master makes a servant do his bidding? If so, he again personifies the purpose itself, and must therefore embody it, or be talking in a manner which plain people cannot understand. If, on the other hand, he means "how simple is the view that the bird acts unconsciously," this is not more simple than supposing it to act consciously; and what ground has he for supposing that the bird is unconscious? It is as simple, and as much in accordance with the facts, to suppose that the bird feels the air to be colder, and knows that she must warm her eggs if she is to hatch them, as consciously as a mother knows that she must not expose her new-born infant to the cold.

On page 99 of this book we find Von Hartmann saying that if it is once granted that the normal and abnormal manifestations of instinct spring from a single source, then the objection that the modification is due to conscious knowledge will be found to be a suicidal one later on, in so far as it is directed against instinct generally. I understand him to mean that if we admit instinctive action, and the modifications of that action which more nearly resemble results of reason, to be actions of the same ultimate kind differing in degree only, and if we thus attempt to reduce instinctive action to the prophetic strain arising from old experience, we shall be obliged to admit that the formation of the embryo is ultimately due to reflection--which he seems to think is a reductio ad absurdum of the argument.

Therefore, he concludes, if there is to be only one source, the source must be unconscious, and not conscious. We reply, that we do not see the absurdity of the position which we grant we have been driven to. We hold that the formation of the embryo IS ultimately due to reflection and design.

The writer of an article in the Times, April 1, 1880, says that servants must be taught their calling before they can practise it; but, in fact, they can only be taught their calling by practising it. So Von Hartmann says animals must feel the pleasure consequent on gratification of an instinct before they can be stimulated to act upon the instinct by a knowledge of the pleasure that will ensue. This sounds logical, but in practice a little performance and a little teaching--a little sense of pleasure and a little connection of that pleasure with this or that practice,--come up simultaneously from something that we cannot see, the two being so small and so much abreast, that we do not know which is first, performance or teaching; and, again, action, or pleasure supposed as coming from the action.

"Geistes-mechanismus" comes as near to "disposition of mind," or, more shortly, "disposition," as so unsatisfactory a word can come to anything. Yet, if we translate it throughout by "disposition," we shall see how little we are being told.

We find on page 114 that "all instinctive actions give us an impression of absolute security and infallibility"; that "the will is never weak or hesitating, as it is when inferences are being drawn consciously." "We never," Von Hartmann continues, "find instinct making mistakes." Passing over the fact that instinct is again personified, the statement is still incorrect. Instinctive actions are certainly, as a general rule, performed with less uncertainty than deliberative ones; this is explicable by the fact that they have been more often practised, and thus reduced more completely to a matter of routine; but nothing is more certain than that animals acting under the guidance of inherited experience or instinct frequently make mistakes which with further practice they correct. Von Hartmann has abundantly admitted that the manner of an instinctive action is often varied in correspondence with variation in external circumstances. It is impossible to see how this does not involve both possibility of error and the connection of instinct with deliberation at one and the same time. The fact is simply this--when an animal finds itself in a like position with that in which it has already often done a certain thing in the persons of its forefathers, it will do this thing well and easily: when it finds the position somewhat, but not unrecognisably, altered through change either in its own person or in the circumstances exterior to it, it will vary its action with greater or less ease according to the nature of the change in the position: when the position is gravely altered the animal either bungles or is completely thwarted.

Not only does Von Hartmann suppose that instinct may, and does, involve knowledge antecedent to, and independent of, experience--an idea as contrary to the tendency of modern thought as that of spontaneous generation, with which indeed it is identical though presented in another shape--but he implies by his frequent use of the word "unmittelbar" that a result can come about without any cause whatever. So he says, "Um fur die unbewusster Erkenntniss, welche nicht durch sinnliche Wahrnehmung erworben, sondern als unmittelbar Besitz," &c. {144a} Because he does not see where the experience can have been gained, he cuts the knot, and denies that there has been experience. We say, Look more attentively and you will discover the time and manner in which the experience was gained.

Again, he continually assumes that animals low down in the scale of life cannot know their own business because they show no sign of knowing ours. See his remarks on Saturnia pavonia minor (page 107), and elsewhere on cattle and gadflies. The question is not what can they know, but what does their action prove to us that they do know. With each species of animal or plant there is one profession only, and it is hereditary. With us there are many professions, and they are not hereditary; so that they cannot become instinctive, as they would otherwise tend to do.

He attempts {144b} to draw a distinction between the causes that have produced the weapons and working instruments of animals, on the one hand, and those that lead to the formation of hexagonal cells by bees, &c., on the other. No such distinction can be justly drawn.

The ghost-stories which Von Hartmann accepts will hardly be accepted by people of sound judgment. There is one well-marked distinctive feature between the knowledge manifested by animals when acting instinctively and the supposed knowledge of seers and clairvoyants. In the first case, the animal never exhibits knowledge except upon matters concerning which its race has been conversant for generations; in the second, the seer is supposed to do so. In the first case, a new feature is invariably attended with disturbance of the performance and the awakening of consciousness and deliberation, unless the new matter is too small in proportion to the remaining features of the case to attract attention, or unless, though really new, it appears so similar to an old feature as to be at first mistaken for it; with the second, it is not even professed that the seer's ancestors have had long experience upon the matter concerning which the seer is supposed to have special insight, and I can imagine no more powerful a priori argument against a belief in such stories.

Close upon the end of his chapter Von Hartmann touches upon the one matter which requires consideration. He refers the similarity of instinct that is observable among all species to the fact that like causes produce like effects; and I gather, though he does not expressly say so, that he considers similarity of instinct in

successive generations to be referable to the same cause as
similarity of instinct between all the contemporary members of a
species. He thus raises the one objection against referring the
phenomena of heredity to memory which I think need be gone into with
any fulness. I will, however, reserve this matter for my concluding
chapters.

Von Hartmann concludes his chapter with a quotation from Schelling,
to the effect that the phenomena of animal instinct are the true
touchstone of a durable philosophy; by which I suppose it is intended
to say that if a system or theory deals satisfactorily with animal
instinct, it will stand, but not otherwise. I can wish nothing
better than that the philosophy of the unconscious advanced by Von
Hartmann be tested by this standard.

CHAPTER X

Recapitulation and statement of an objection.

The true theory of unconscious action, then, is that of Professor
Hering, from whose lecture it is no strained conclusion to gather
that he holds the action of all living beings, from the moment of
their conception to that of their fullest development, to be founded
in volition and design, though these have been so long lost sight of
that the work is now carried on, as it were, departmentally and in
due course according to an official routine which can hardly now be
departed from.

This involves the older "Darwinism" and the theory of Lamarck, according to which the modification of living forms has been effected mainly through the needs of the living forms themselves, which vary with varying conditions, the survival of the fittest (which, as I see Mr. H. B. Baildon has just said, "sometimes comes to mean merely the survival of the survivors" {146}) being taken almost as a matter of course. According to this view of evolution, there is a remarkable analogy between the development of living organs or tools and that of those organs or tools external to the body which has been so rapid during the last few thousand years.

Animals and plants, according to Professor Hering, are guided throughout their development, and preserve the due order in each step which they take, through memory of the course they took on past occasions when in the persons of their ancestors. I am afraid I have already too often said that if this memory remains for long periods together latent and without effect, it is because the undulations of the molecular substance of the body which are its supposed explanation are during these periods too feeble to generate action, until they are augmented in force through an accession of suitable undulations issuing from exterior objects; or, in other words, until recollection is stimulated by a return of the associated ideas. On this the eternal agitation becomes so much enhanced, that equilibrium is visibly disturbed, and the action ensues which is proper to the vibration of the particular substance under the particular conditions. This, at least, is what I suppose Professor Hering to intend.

Leaving the explanation of memory on one side, and confining ourselves to the fact of memory only, a caterpillar on being just hatched is supposed, according to this theory, to lose its memory of the time it was in the egg, and to be stimulated by an intense but unconscious recollection of the action taken by its ancestors when they were first hatched. It is guided in the course it takes by the experience it can thus command. Each step it takes recalls a new recollection, and thus it goes through its development as a performer performs a piece of music, each bar leading his recollection to the bar that should next follow.

In "Life and Habit" will be found examples of the manner in which this view solves a number of difficulties for the explanation of which the leading men of science express themselves at a loss. The following from Professor Huxley's recent work upon the crayfish may serve for an example. Professor Huxley writes:-

"It is a widely received notion that the energies of living matter have a tendency to decline and finally disappear, and that the death of the body as a whole is a necessary correlate of its life. That all living beings sooner or later perish needs no demonstration, but it would be difficult to find satisfactory grounds for the belief that they needs must do so. The analogy of a machine, that sooner or later must be brought to a standstill by the wear and tear of its parts, does not hold, inasmuch as the animal mechanism is continually renewed and repaired; and though it is true that individual components of the body are constantly dying, yet their places are taken by vigorous successors. A city remains notwithstanding the constant death-rate of its inhabitants; and such an organism as a crayfish is only a corporate unity, made up of innumerable partially independent individualities."--The Crayfish, p. 127.

Surely the theory which I have indicated above makes the reason plain why no organism can permanently outlive its experience of past lives. The death of such a body corporate as the crayfish is due to the social condition becoming more complex than there is memory of past experience to deal with. Hence social disruption, insubordination, and decay. The crayfish dies as a state dies, and all states that we have heard of die sooner or later. There are some savages who have not yet arrived at the conception that death is the necessary end of all living beings, and who consider even the gentlest death from old age as violent and abnormal; so Professor Huxley seems to find a difficulty in seeing that though a city commonly outlives many generations of its citizens, yet cities and states are in the end no less mortal than individuals. "The city," he says, "remains." Yes, but not for ever. When Professor Huxley can find a city that will last for ever, he may wonder that a crayfish does not last for ever.

I have already here and elsewhere said all that I can yet bring forward in support of Professor Hering's theory; it now remains for me to meet the most troublesome objection to it that I have been able to think of--an objection which I had before me when I wrote "Life and Habit," but which then as now I believe to be unsound. Seeing, however, as I have pointed out at the end of the preceding chapter, that Von Hartmann has touched upon it, and being aware that a plausible case can be made out for it, I will state it and refute it here. When I say refute it, I do not mean that I shall have done with it--for it is plain that it opens up a vaster question in the relations between the so-called organic and inorganic worlds--but that I will refute the supposition that it any way militates against Professor Hering's theory.

Why, it may be asked, should we go out of our way to invent unconscious memory--the existence of which must at the best remain an inference {149}--when the observed fact that like antecedents are invariably followed by like consequents should be sufficient for our purpose? Why should the fact that a given kind of chrysalis in a given condition will always become a butterfly within a certain time be connected with memory, when it is not pretended that memory has anything to do with the invariableness with which oxygen and hydrogen when mixed in certain proportions make water?

We assume confidently that if a drop of water were decomposed into its component parts, and if these were brought together again, and again decomposed and again brought together any number of times over, the results would be invariably the same, whether decomposition or combination, yet no one will refer the invariableness of the action during each repetition, to recollection by the gaseous molecules of the course taken when the process was last repeated. On the contrary, we are assured that molecules in some distant part of the world, which had never entered into such and such a known combination themselves, nor held concert with other molecules that had been so combined, and which, therefore, could have had no experience and no memory, would none the less act upon one another in that one way in which other like combinations of atoms have acted under like circumstances, as readily as though they had been combined and separated and recombined again a hundred or a hundred thousand times. It is this assumption, tacitly made by every man, beast, and plant in the universe, throughout all time and in every action of their lives, that has made any action possible, lying, as it does, at the root of all experience.

As we admit of no doubt concerning the main result, so we do not suppose an alternative to lie before any atom of any molecule at any moment during the process of their combination. This process is, in all probability, an exceedingly complicated one, involving a multitude of actions and subordinate processes, which follow one upon the other, and each one of which has a beginning, a middle, and an end, though they all come to pass in what appears to be an instant of time. Yet at no point do we conceive of any atom as swerving ever such a little to right or left of a determined course, but invest each one of them with so much of the divine attributes as that with it there shall be no variableness, neither shadow of turning.

We attribute this regularity of action to what we call the necessity of things, as determined by the nature of the atoms and the circumstances in which they are placed. We say that only one proximate result can ever arise from any given combination. If, then, so great uniformity of action as nothing can exceed is manifested by atoms to which no one will impute memory, why this desire for memory, as though it were the only way of accounting for regularity of action in living beings? Sameness of action may be seen abundantly where there is no room for anything that we can consistently call memory. In these cases we say that it is due to sameness of substance in same circumstances.

The most cursory reflection upon our actions will show us that it is no more possible for living action to have more than one set of proximate consequents at any given time than for oxygen and hydrogen when mixed in the proportions proper for the formation of water. Why, then, not recognise this fact, and ascribe repeated similarity of living action to the reproduction of the necessary antecedents, with no more sense of connection between the steps in the action, or memory of similar action taken before, than we suppose on the part of oxygen and hydrogen molecules between the several occasions on which they may have been disunited and reunited?

A boy catches the measles not because he remembers having caught them in the persons of his father and mother, but because he is a fit soil for a certain kind of seed to grow upon. In like manner he should be said to grow his nose because he is a fit combination for a nose to spring from. Dr. X---'s father died of angina pectoris at the age of forty-nine; so did Dr. X---. Can it be pretended that Dr. X--- remembered having died of angina pectoris at the age of forty-nine when in the person of his father, and accordingly, when he came to be forty-nine years old himself, died also? For this to hold, Dr. X---

's father must have begotten him after he was dead; for the son could not remember the father's death before it happened.

As for the diseases of old age, so very commonly inherited, they are developed for the most part not only long after the average age of reproduction, but at a time when no appreciable amount of memory of any previous existence can remain; for a man will not have many male ancestors who become parents at over sixty years old, nor female ancestors who did so at over forty. By our own showing, therefore, recollection can have nothing to do with the matter. Yet who can doubt that gout is due to inheritance as much as eyes and noses? In what respects do the two things differ so that we should refer the inheritance of eyes and noses to memory, while denying any connection between memory and gout? We may have a ghost of a pretence for saying that a man grew a nose by rote, or even that he catches the measles or whooping-cough by rote during his boyhood; but do we mean to say that he develops the gout by rote in his old age if he comes of a gouty family? If, then, rote and red-tape have nothing to do with the one, why should they with the other?

Remember also the cases in which aged females develop male characteristics. Here are growths, often of not inconsiderable extent, which make their appearance during the decay of the body, and grow with greater and greater vigour in the extreme of old age, and even for days after death itself. It can hardly be doubted that an especial tendency to develop these characteristics runs as an inheritance in certain families; here then is perhaps the best case that can be found of a development strictly inherited, but having clearly nothing whatever to do with memory. Why should not all development stand upon the same footing?

A friend who had been arguing with me for some time as above, concluded with the following words:-

"If you cannot be content with the similar action of similar substances (living or non-living) under similar circumstances--if you cannot accept this as an ultimate fact, but consider it necessary to connect repetition of similar action with memory before you can rest in it and be thankful--be consistent, and introduce this memory which you find so necessary into the inorganic world also. Either say that a chrysalis becomes a butterfly because it is the thing that it is, and, being that kind of thing, must act in such and such a manner and in such a manner only, so that the act of one generation has no more to do with the act of the next than the fact of cream being churned into butter in a dairy one day has to do with other cream being churnable into butter in the following week--either say this, or else develop some mental condition--which I have no doubt you will be very well able to do if you feel the want of it--in which you can make out a case for saying that oxygen and hydrogen on being brought together, and cream on being churned, are in some way acquainted with, and mindful of, action taken by other cream and other oxygen and hydrogen on past occasions."

I felt inclined to reply that my friend need not twit me with being able to develop a mental organism if I felt the need of it, for his own ingenious attack on my position, and indeed every action of his life was but an example of this omnipresent principle.

When he was gone, however, I thought over what he had been saying. I endeavoured to see how far I could get on without volition and memory, and reasoned as follows:- A repetition of like antecedents will be certainly followed by a repetition of like consequents, whether the agents be men and women or chemical substances. "If there be two cowards perfectly similar in every respect, and if they be subjected in a perfectly similar way to two terrifying agents, which are themselves perfectly similar, there are few who will not expect a perfect similarity in the running away, even though ten thousand years intervene between the original combination and its repetition." {153} Here certainly there is no coming into play of memory, more than in the pan of cream on two successive churning days, yet the action is similar.

A clerk in an office has an hour in the middle of the day for dinner. About half-past twelve he begins to feel hungry; at once he takes down his hat and leaves the office. He does not yet know the neighbourhood, and on getting down into the street asks a policeman at the corner which is the best eating-house within easy distance. The policeman tells him of three houses, one of which is a little farther off than the other two, but is cheaper. Money being a greater object to him than time, the clerk decides on going to the cheaper house. He goes, is satisfied, and returns.

Next day he wants his dinner at the same hour, and--it will be said-- remembering his satisfaction of yesterday, will go to the same place as before. But what has his memory to do with it? Suppose him to have entirely forgotten all the circumstances of the preceding day from the moment of his beginning to feel hungry onward, though in other respects sound in mind and body, and unchanged generally. At half-past twelve he would begin to be hungry; but his beginning to be hungry cannot be connected with his remembering having begun to be hungry yesterday. He would begin to be hungry just as much whether he remembered or no. At one o'clock he again takes down his hat and leaves the office, not because he remembers having done so yesterday, but because he wants his hat to go out with. Being again in the street, and again ignorant of the neighbourhood (for he remembers nothing of yesterday), he sees the same policeman at the corner of the street, and asks him the same question as before; the policeman gives him the same answer, and money being still an object to him, the cheapest eating-house is again selected; he goes there, finds the same menu, makes the same choice for the same reasons, eats, is satisfied, and returns.

What similarity of action can be greater than this, and at the same time more incontrovertible? But it has nothing to do with memory; on the contrary, it is just because the clerk has no memory that his action of the second day so exactly resembles that of the first. As long as he has no power of recollecting, he will day after day repeat the same actions in exactly the same way, until some external circumstances, such as his being sent away, modify the situation. Till this or some other modification occurs, he will day after day go down into the street without knowing where to go; day after day he will see the same policeman at the corner of the same street, and (for we may as well suppose that the policeman has no memory too) he will ask and be answered, and ask and be answered, till he and the policeman die of old age. This similarity of action is plainly due to that--whatever it is--which ensures that like persons or things when placed in like circumstances shall behave in like manner.

Allow the clerk ever such a little memory, and the similarity of action will disappear; for the fact of remembering what happened to him on the first day he went out in search of dinner will be a modification in him in regard to his then condition when he next goes out to get his dinner. He had no such memory on the first day, and he has upon the second. Some modification of action must ensue upon this modification of the actor, and this is immediately observable. He wants his dinner, indeed, goes down into the street, and sees the policeman as yesterday, but he does not ask the policeman; he remembers what the policeman told him and what he did, and therefore goes straight to the eating-house without wasting time: nor does he dine off the same dish two days running, for he remembers what he had yesterday and likes variety. If, then, similarity of action is rather hindered than promoted by memory, why introduce it into such cases as the repetition of the embryonic processes by successive generations? The embryos of a well-fixed breed, such as the goose, are almost as much alike as water is to water, and by consequence one goose comes to be almost as like another as water to water. Why should it not be supposed to become so upon the same grounds--namely, that it is made of the same stuffs, and put together in like proportions in the same manner?

CHAPTER XI

On Cycles.

The one faith on which all normal living beings consciously or unconsciously act, is that like antecedents will be followed by like consequents. This is the one true and catholic faith, undemonstrable, but except a living being believe which, without doubt it shall perish everlastingly. In the assurance of this all action is taken.

But if this fundamental article is admitted, and it cannot be gainsaid, it follows that if ever a complete cycle were formed, so that the whole universe of one instant were to repeat itself absolutely in a subsequent one, no matter after what interval of time, then the course of the events between these two moments would go on repeating itself for ever and ever afterwards in due order, down to the minutest detail, in an endless series of cycles like a circulating decimal. For the universe comprises everything; there could therefore be no disturbance from without. Once a cycle, always a cycle.

Let us suppose the earth, of given weight, moving with given momentum in a given path, and under given conditions in every respect, to find itself at any one time conditioned in all these respects as it was conditioned at some past moment; then it must move exactly in the same path as the one it took when at the beginning of the cycle it has just completed, and must therefore in the course of time fulfil a second cycle, and therefore a third, and so on for ever and ever, with no more chance of escape than a circulating decimal has, if the circumstances have been reproduced with perfect accuracy.

We see something very like this actually happen in the yearly revolutions of the planets round the sun. But the relations between, we will say, the earth and the sun are not reproduced absolutely. These relations deal only with a small part of the universe, and even in this small part the relation of the parts inter se has never yet been reproduced with the perfection of accuracy necessary for our argument. They are liable, moreover, to disturbance from events which may or may not actually occur (as, for example, our being struck by a comet, or the sun's coming within a certain distance of another sun), but of which, if they do occur, no one can foresee the effects. Nevertheless the conditions have been so nearly repeated that there is no appreciable difference in the relations between the earth and sun on one New Year's Day and on another, nor is there reason for expecting such change within any reasonable time.

If there is to be an eternal series of cycles involving the whole universe, it is plain that not one single atom must be excluded. Exclude a single molecule of hydrogen from the ring, or vary the relative positions of two molecules only, and the charm is broken; an element of disturbance has been introduced, of which the utmost that can be said is that it may not prevent the ensuing of a long series of very nearly perfect cycles before similarity in recurrence is destroyed, but which must inevitably prevent absolute identity of repetition. The movement of the series becomes no longer a cycle, but spiral, and convergent or divergent at a greater or less rate according to circumstances. We cannot conceive of all the atoms in the universe standing twice over in absolutely the same relation each one of them to every other. There are too many of them and they are too much mixed; but, as has been just said, in the planets and their satellites we do see large groups of atoms whose movements recur with some approach to precision. The same holds good also with certain comets and with the sun himself. The result is that our days and nights and seasons follow one another with nearly perfect regularity from year to year, and have done so for as long time as we know anything for certain. A vast preponderance of all the action that takes place around us is cycular action.

Within the great cycle of the planetary revolution of our own earth, and as a consequence thereof, we have the minor cycle of the phenomena of the seasons; these generate atmospheric cycles. Water is evaporated from the ocean and conveyed to mountain ranges, where it is cooled, and whence it returns again to the sea. This cycle of events is being repeated again and again with little appreciable variation. The tides and winds in certain latitudes go round and round the world with what amounts to continuous regularity.--There are storms of wind and rain called cyclones. In the case of these, the cycle is not very complete, the movement, therefore, is spiral, and the tendency to recur is comparatively soon lost. It is a common saying that history repeats itself, so that anarchy will lead to despotism and despotism to anarchy; every nation can point to instances of men's minds having gone round and round so nearly in a perfect cycle that many revolutions have occurred before the cessation of a tendency to recur. Lastly, in the generation of plants and animals we have, perhaps, the most striking and common example of the inevitable tendency of all action to repeat itself when it has once proximately done so. Let only one living being have once succeeded in producing a being like itself, and thus have returned, so to speak, upon itself, and a series of generations must follow of necessity, unless some matter interfere which had no part in the original combination, and, as it may happen, kill the first reproductive creature or all its descendants within a few generations. If no such mishap occurs as this, and if the recurrence of the conditions is sufficiently perfect, a series of generations follows with as much certainty as a series of seasons follows upon the cycle of the relations between the earth and sun. Let the first periodically recurring substance--we will say A--be able to recur or reproduce itself, not once only, but many times over, as A1, A2, &c.; let A also have consciousness and a sense of self-interest, which qualities must, ex hypothesi, be reproduced in each one of its offspring; let these get placed in circumstances which differ sufficiently to destroy the cycle in theory without doing so practically--that is to say, to reduce the rotation to a spiral, but to a spiral with so little deviation from perfect cycularity as for each revolution to appear practically a cycle, though after many revolutions the deviation becomes perceptible; then some such differentiations of animal and vegetable life as we actually see follow as matters of course. A1 and A2 have a sense of self-interest as A had, but they are not precisely in circumstances similar to A's, nor, it may be, to each other's; they will therefore act somewhat differently, and every living being is modified by a change of action. Having become modified, they follow the spirit of A's action more essentially in begetting a creature like themselves than in begetting one like A; for the essence of A's act was not the reproduction of A, but the reproduction of a creature like the one from which it sprung--that is to say, a creature bearing traces in its body of the main influences that have worked upon its parent.

Within the cycle of reproduction there are cycles upon cycles in the

life of each individual, whether animal or plant. Observe the action
of our lungs and heart, how regular it is, and how a cycle having
been once established, it is repeated many millions of times in an
individual of average health and longevity. Remember also that it is
this periodicity--this inevitable tendency of all atoms in
combination to repeat any combination which they have once repeated,
unless forcibly prevented from doing so--which alone renders nine-
tenths of our mechanical inventions of practical use to us. There is
no internal periodicity about a hammer or a saw, but there is in the
steam-engine or watermill when once set in motion. The actions of
these machines recur in a regular series, at regular intervals, with
the unerringness of circulating decimals.

When we bear in mind, then, the omnipresence of this tendency in the
world around us, the absolute freedom from exception which attends
its action, the manner in which it holds equally good upon the
vastest and the smallest scale, and the completeness of its accord
with our ideas of what must inevitably happen when a like combination
is placed in circumstances like those in which it was placed before--
when we bear in mind all this, is it possible not to connect the
facts together, and to refer cycles of living generations to the same
unalterableness in the action of like matter under like circumstances
which makes Jupiter and Saturn revolve round the sun, or the piston
of a steam-engine move up and down as long as the steam acts upon it?

But who will attribute memory to the hands of a clock, to a piston-
rod, to air or water in a storm or in course of evaporation, to the
earth and planets in their circuits round the sun, or to the atoms of
the universe, if they too be moving in a cycle vaster than we can
take account of? {160} And if not, why introduce it into the
embryonic development of living beings, when there is not a particle
of evidence in support of its actual presence, when regularity of
action can be ensured just as well without it as with it, and when at

the best it is considered as existing under circumstances which it baffles us to conceive, inasmuch as it is supposed to be exercised without any conscious recollection? Surely a memory which is exercised without any consciousness of recollecting is only a periphrasis for the absence of any memory at all.

CHAPTER XII

Refutation--Memory at once a promoter and a disturber of uniformity of action and structure.

To meet the objections in the two foregoing chapters, I need do little more than show that the fact of certain often inherited diseases and developments, whether of youth or old age, being obviously not due to a memory on the part of offspring of like diseases and developments in the parents, does not militate against supposing that embryonic and youthful development generally is due to memory.

This is the main part of the objection; the rest resolves itself into an assertion that there is no evidence in support of instinct and embryonic development being due to memory, and a contention that the necessity of each particular moment in each particular case is sufficient to account for the facts without the introduction of memory.

I will deal with these two last points briefly first. As regards the evidence in support of the theory that instinct and growth are due to a rapid unconscious memory of past experiences and developments in

the persons of the ancestors of the living form in which they appear,
I must refer my readers to "Life and Habit," and to the translation
of Professor Hering's lecture given in this volume. I will only
repeat here that a chrysalis, we will say, is as much one and the
same person with the chrysalis of its preceding generation, as this
last is one and the same person with the egg or caterpillar from
which it sprang. You cannot deny personal identity between two
successive generations without sooner or later denying it during the
successive stages in the single life of what we call one individual;
nor can you admit personal identity through the stages of a long and
varied life (embryonic and postnatal) without admitting it to endure
through an endless series of generations.

The personal identity of successive generations being admitted, the
possibility of the second of two generations remembering what
happened to it in the first is obvious. The a priori objection,
therefore, is removed, and the question becomes one of fact--does the
offspring act as if it remembered?

The answer to this question is not only that it does so act, but that
it is not possible to account for either its development or its early
instinctive actions upon any other hypothesis than that of its
remembering, and remembering exceedingly well.

The only alternative is to declare with Von Hartmann that a living
being may display a vast and varied information concerning all manner
of details, and be able to perform most intricate operations,
independently of experience and practice. Once admit knowledge
independent of experience, and farewell to sober sense and reason
from that moment.

Firstly, then, we show that offspring has had every facility for
remembering; secondly, that it shows every appearance of having

remembered; thirdly, that no other hypothesis except memory can be brought forward, so as to account for the phenomena of instinct and heredity generally, which is not easily reducible to an absurdity. Beyond this we do not care to go, and must allow those to differ from us who require further evidence.

As regards the argument that the necessity of each moment will account for likeness of result, without there being any need for introducing memory, I admit that likeness of consequents is due to likeness of antecedents, and I grant this will hold as good with embryos as with oxygen and hydrogen gas; what will cover the one will cover the other, for time writs of the laws common to all matter run within the womb as freely as elsewhere; but admitting that there are combinations into which living beings enter with a faculty called memory which has its effect upon their conduct, and admitting that such combinations are from time to time repeated (as we observe in the case of a practised performer playing a piece of music which he has committed to memory), then I maintain that though, indeed, the likeness of one performance to its immediate predecessor is due to likeness of the combinations immediately preceding the two performances, yet memory plays so important a part in both these combinations as to make it a distinguishing feature in them, and therefore proper to be insisted upon. We do not, for example, say that Herr Joachim played such and such a sonata without the music, because he was such and such an arrangement of matter in such and such circumstances, resembling those under which he played without music on some past occasion. This goes without saying; we say only that he played the music by heart or by memory, as he had often played it before.

To the objector that a caterpillar becomes a chrysalis not because it remembers and takes the action taken by its fathers and mothers in due course before it, but because when matter is in such a physical and mental state as to be called caterpillar, it must perforce assume presently such another physical and mental state as to be called chrysalis, and that therefore there is no memory in the case--to this objector I rejoin that the offspring caterpillar would not have become so like the parent as to make the next or chrysalis stage a matter of necessity, unless both parent and offspring had been influenced by something that we usually call memory. For it is this very possession of a common memory which has guided the offspring into the path taken by, and hence to a virtually same condition with, the parent, and which guided the parent in its turn to a state virtually identical with a corresponding state in the existence of its own parent. To memory, therefore, the most prominent place in the transaction is assigned rightly.

To deny that will guided by memory has anything to do with the development of embryos seems like denying that a desire to obstruct has anything to do with the recent conduct of certain members in the House of Commons. What should we think of one who said that the action of these gentlemen had nothing to do with a desire to embarrass the Government, but was simply the necessary outcome of the chemical and mechanical forces at work, which being such and such, the action which we see is inevitable, and has therefore nothing to do with wilful obstruction? We should answer that there was doubtless a great deal of chemical and mechanical action in the matter; perhaps, for aught we knew or cared, it was all chemical and mechanical; but if so, then a desire to obstruct parliamentary business is involved in certain kinds of chemical and mechanical action, and that the kinds involving this had preceded the recent proceedings of the members in question. If asked to prove this, we can get no further than that such action as has been taken has never yet been seen except as following after and in consequence of a desire to obstruct; that this is our nomenclature, and that we can no more be expected to change it than to change our mother tongue at the bidding of a foreigner.

A little reflection will convince the reader that he will be unable
to deny will and memory to the embryo without at the same time
denying their existence everywhere, and maintaining that they have no
place in the acquisition of a habit, nor indeed in any human action.
He will feel that the actions, and the relation of one action to
another which he observes in embryos is such as is never seen except
in association with and as a consequence of will and memory. He will
therefore say that it is due to will and memory. To say that these
are the necessary outcome of certain antecedents is not to destroy
them: granted that they are--a man does not cease to be a man when
we reflect that he has had a father and mother, nor do will and
memory cease to be will and memory on the ground that they cannot
come causeless. They are manifest minute by minute to the perception
of all sane people, and this tribunal, though not infallible, is
nevertheless our ultimate court of appeal--the final arbitrator in
all disputed cases.

We must remember that there is no action, however original or
peculiar, which is not in respect of far the greater number of its
details founded upon memory. If a desperate man blows his brains
out--an action which he can do once in a lifetime only, and which
none of his ancestors can have done before leaving offspring--still
nine hundred and ninety-nine thousandths of the movements necessary
to achieve his end consist of habitual movements--movements, that is
to say, which were once difficult, but which have been practised and
practised by the help of memory until they are now performed
automatically. We can no more have an action than a creative effort
of the imagination cut off from memory. Ideas and actions seem
almost to resemble matter and force in respect of the impossibility
of originating or destroying them; nearly all that are, are memories
of other ideas and actions, transmitted but not created, disappearing

but not perishing.

It appears, then, that when in Chapter X. we supposed the clerk who wanted his dinner to forget on a second day the action he had taken the day before, we still, without perhaps perceiving it, supposed him to be guided by memory in all the details of his action, such as his taking down his hat and going out into the street. We could not, indeed, deprive him of all memory without absolutely paralysing his action.

Nevertheless new ideas, new faiths, and new actions do in the course of time come about, the living expressions of which we may see in the new forms of life which from time to time have arisen and are still arising, and in the increase of our own knowledge and mechanical inventions. But it is only a very little new that is added at a time, and that little is generally due to the desire to attain an end which cannot be attained by any of the means for which there exists a perceived precedent in the memory. When this is the case, either the memory is further ransacked for any forgotten shreds of details, a combination of which may serve the desired purpose; or action is taken in the dark, which sometimes succeeds and becomes a fertile source of further combinations; or we are brought to a dead stop. All action is random in respect of any of the minute actions which compose it that are not done in consequence of memory, real or supposed. So that random, or action taken in the dark, or illusion, lies at the very root of progress.

I will now consider the objection that the phenomena of instinct and embryonic development ought not to be ascribed to memory, inasmuch as certain other phenomena of heredity, such as gout, cannot be ascribed to it.

Those who object in this way forget that our actions fall into two main classes: those which we have often repeated before by means of a regular series of subordinate actions beginning and ending at a certain tolerably well-defined point--as when Herr Joachim plays a sonata in public, or when we dress or undress ourselves; and actions the details of which are indeed guided by memory, but which in their general scope and purpose are new--as when we are being married or presented at court.

At each point in any action of the first of the two kinds above referred to there is a memory (conscious or unconscious according to the less or greater number of times the action has been repeated), not only of the steps in the present and previous performances which have led up to the particular point that may be selected, but also of the particular point itself; there is, therefore, at each point in a habitual performance a memory at once of like antecedents and of a like present.

If the memory, whether of the antecedent or the present, were absolutely perfect; if the vibration (according to Professor Hering) on each repetition existed in its full original strength and without having been interfered with by any other vibration; and if, again, the new wave running into it from exterior objects on each repetition of the action were absolutely identical in character with the wave that ran in upon the last occasion, then there would be no change in the action and no modification or improvement could take place. For though indeed the latest performance would always have one memory more than the latest but one to guide it, yet the memories being identical, it would not matter how many or how few they were.

On any repetition, however, the circumstances, external or internal, or both, never are absolutely identical: there is some slight variation in each individual case, and some part of this variation is remembered, with approbation or disapprobation as the case may be.

The fact, therefore, that on each repetition of the action there is one memory more than on the last but one, and that this memory is slightly different from its predecessor, is seen to be an inherent and, ex hypothesi, necessarily disturbing factor in all habitual action--and the life of an organism should be regarded as the habitual action of a single individual, namely, of the organism itself, and of its ancestors. This is the key to accumulation of improvement, whether in the arts which we assiduously practise during our single life, or in the structures and instincts of successive generations. The memory does not complete a true circle, but is, as it were, a spiral slightly divergent therefrom. It is no longer a perfectly circulating decimal. Where, on the other hand, there is no memory of a like present, where, in fact, the memory is not, so to speak, spiral, there is no accumulation of improvement. The effect of any variation is not transmitted, and is not thus pregnant of still further change.

As regards the second of the two classes of actions above referred to--those, namely, which are not recurrent or habitual, AND AT NO POINT OF WHICH IS THERE A MEMORY OF A PAST PRESENT LIKE THE ONE WHICH
IS PRESENT NOW--there will have been no accumulation of strong and well-knit memory as regards the action as a whole, but action, if taken at all, will be taken upon disjointed fragments of individual actions (our own and those of other people) pieced together with a result more or less satisfactory according to circumstances.

But it does not follow that the action of two people who have had tolerably similar antecedents and are placed in tolerably similar circumstances should be more unlike each other in this second case than in the first. On the contrary, nothing is more common than to observe the same kind of people making the same kind of mistake when placed for the first time in the same kind of new circumstances. I did not say that there would be no sameness of action without memory of a like present. There may be sameness of action proceeding from a memory, conscious or unconscious, of like antecedents, and A PRESENCE ONLY OF LIKE PRESENTS WITHOUT RECOLLECTION OF THE SAME.

The sameness of action of like persons placed under like circumstances for the first time, resembles the sameness of action of inorganic matter under the same combinations. Let us for the moment suppose what we call non-living substances to be capable of remembering their antecedents, and that the changes they undergo are the expressions of their recollections. Then I admit, of course, that there is not memory in any cream, we will say, that is about to be churned of the cream of the preceding week, but the common absence of such memory from each week's cream is an element of sameness between the two. And though no cream can remember having been churned before, yet all cream in all time has had nearly identical antecedents, and has therefore nearly the same memories, and nearly the same proclivities. Thus, in fact, the cream of one week is as truly the same as the cream of another week from the same cow, pasture, &c., as anything is ever the same with anything; for the having been subjected to like antecedents engenders the closest similarity that we can conceive of, if the substances were like to start with.

The manifest absence of any connecting memory (or memory of like presents) from certain of the phenomena of heredity, such as, for example, the diseases of old age, is now seen to be no valid reason for saying that such other and far more numerous and important phenomena as those of embryonic development are not phenomena of memory. Growth and the diseases of old age do indeed, at first sight, appear to stand on the same footing, but reflection shows us that the question whether a certain result is due to memory or no

must be settled not by showing that combinations into which memory
does not certainly enter may yet generate like results, and therefore
considering the memory theory disposed of, but by the evidence we may
be able to adduce in support of the fact that the second agent has
actually remembered the conduct of the first, inasmuch as he cannot
be supposed able to do what it is plain he can do, except under the
guidance of memory or experience, and can also be shown to have had
every opportunity of remembering. When either of these tests fails,
similarity of action on the part of two agents need not be connected
with memory of a like present as well as of like antecedents, but
must, or at any rate may, be referred to memory of like antecedents
only.

Returning to a parenthesis a few pages back, in which I said that
consciousness of memory would be less or greater according to the
greater or fewer number of times that the act had been repeated, it
may be observed as a corollary to this, that the less consciousness
of memory the greater the uniformity of action, and vice versa. For
the less consciousness involves the memory's being more perfect,
through a larger number (generally) of repetitions of the act that is
remembered; there is therefore a less proportionate difference in
respect of the number of recollections of this particular act between
the most recent actor and the most recent but one. This is why very
old civilisations, as those of many insects, and the greater number
of now living organisms, appear to the eye not to change at all.

For example, if an action has been performed only ten times, we will
say by A, B, C, &c., who are similar in all respects, except that A
acts without recollection, B with recollection of A's action, C with
recollection of both B's and A's, while J remembers the course taken
by A, B, C, D, E, F, G, H, and I--the possession of a memory by B
will indeed so change his action, as compared with A's, that it may
well be hardly recognisable. We saw this in our example of the clerk

who asked the policeman the way to the eating-house on one day, but
did not ask him the next, because he remembered; but C's action will
not be so different from B's as B's from A's, for though C will act
with a memory of two occasions on which the action has been
performed, while B recollects only the original performance by A, yet
B and C both act with the guidance of a memory and experience of some
kind, while A acted without any. Thus the clerk referred to in
Chapter X. will act on the third day much as he acted on the second--
that is to say, he will see the policeman at the corner of the
street, but will not question him.

When the action is repeated by J for the tenth time, the difference
between J's repetition of it and I's will be due solely to the
difference between a recollection of nine past performances by J
against only eight by I, and this is so much proportionately less
than the difference between a recollection of two performances and of
only one, that a less modification of action should be expected. At
the same time consciousness concerning an action repeated for the
tenth time should be less acute than on the first repetition.
Memory, therefore, though tending to disturb similarity of action
less and less continually, must always cause some disturbance. At
the same time the possession of a memory on the successive
repetitions of an action after the first, and, perhaps, the first two
or three, during which the recollection may be supposed still
imperfect, will tend to ensure uniformity, for it will be one of the
elements of sameness in the agents--they both acting by the light of
experience and memory.

During the embryonic stages and in childhood we are almost entirely
under the guidance of a practised and powerful memory of
circumstances which have been often repeated, not only in detail and
piecemeal, but as a whole, and under many slightly varying
conditions; thus the performance has become well averaged and matured

in its arrangements, so as to meet all ordinary emergencies. We therefore act with great unconsciousness and vary our performances little. Babies are much more alike than persons of middle age.

Up to the average age at which our ancestors have had children during many generations, we are still guided in great measure by memory; but the variations in external circumstances begin to make themselves perceptible in our characters. In middle life we live more and more continually upon the piecing together of details of memory drawn from our personal experience, that is to say, upon the memory of our own antecedents; and this resembles the kind of memory we hypothetically attached to cream a little time ago. It is not surprising, then, that a son who has inherited his father's tastes and constitution, and who lives much as his father had done, should make the same mistakes as his father did when he reaches his father's age--we will say of seventy--though he cannot possibly remember his father's having made the mistakes. It were to be wished we could, for then we might know better how to avoid gout, cancer, or what not. And it is to be noticed that the developments of old age are generally things we should be glad enough to avoid if we knew how to do so.

CHAPTER XIII

Conclusion.

If we observed the resemblance between successive generations to be as close as that between distilled water and distilled water through all time, and if we observed that perfect unchangeableness in the action of living beings which we see in what we call chemical and mechanical combinations, we might indeed suspect that memory had as little place among the causes of their action as it can have in anything, and that each repetition, whether of a habit or the practice of art, or of an embryonic process in successive generations, was an original performance, for all that memory had to do with it. I submit, however, that in the case of the reproductive forms of life we see just so much variety, in spite of uniformity, as is consistent with a repetition involving not only a nearly perfect similarity in the agents and their circumstances, but also the little departure therefrom that is inevitably involved in the supposition that a memory of like presents as well as of like antecedents (as distinguished from a memory of like antecedents only) has played a part in their development--a cyclonic memory, if the expression may be pardoned.

There is life infinitely lower and more minute than any which our most powerful microscopes reveal to us, but let us leave this upon one side and begin with the amoeba. Let us suppose that this structureless morsel of protoplasm is, for all its structurelessness, composed of an infinite number of living molecules, each one of them with hopes and fears of its own, and all dwelling together like Tekke Turcomans, of whom we read that they live for plunder only, and that each man of them is entirely independent, acknowledging no constituted authority, but that some among them exercise a tacit and undefined influence over the others. Let us suppose these molecules capable of memory, both in their capacity as individuals, and as societies, and able to transmit their memories to their descendants, from the traditions of the dimmest past to the experiences of their own lifetime. Some of these societies will remain simple, as having had no history, but to the greater number unfamiliar, and therefore striking, incidents will from time to time occur, which, when they do not disturb memory so greatly as to kill, will leave their impression upon it. The body or society will remember these incidents, and be modified by them in its conduct, and therefore more or less in its internal arrangements, which will tend inevitably to specialisation. This memory of the most striking events of varied lifetimes I maintain, with Professor Hering, to be the differentiating cause, which, accumulated in countless generations, has led up from the amoeba to man. If there had been no such memory, the amoeba of one generation would have exactly resembled time amoeba of the preceding, and a perfect cycle would have been established; the modifying effects of an additional memory in each generation have made the cycle into a spiral, and into a spiral whose eccentricity, in the outset hardly perceptible, is becoming greater and greater with increasing longevity and more complex social and mechanical inventions.

We say that the chicken grows the horny tip to its beak with which it ultimately pecks its way out of its shell, because it remembers having grown it before, and the use it made of it. We say that it made it on the same principles as a man makes a spade or a hammer, that is to say, as the joint result both of desire and experience. When I say experience, I mean experience not only of what will be wanted, but also of the details of all the means that must be taken in order to effect this. Memory, therefore, is supposed to guide the chicken not only in respect of the main design, but in respect also of every atomic action, so to speak, which goes to make up the execution of this design. It is not only the suggestion of a plan which is due to memory, but, as Professor Hering has so well said, it is the binding power of memory which alone renders any consolidation or coherence of action possible, inasmuch as without this no action could have parts subordinate one to another, yet bearing upon a common end; no part of an action, great or small, could have reference to any other part, much less to a combination of all the parts; nothing, in fact, but ultimate atoms of actions could ever happen--these bearing the same relation to such an action, we will

say, as a railway journey from London to Edinburgh as a single
molecule of hydrogen to a gallon of water. If asked how it is that
the chicken shows no sign of consciousness concerning this design,
nor yet of the steps it is taking to carry it out, we reply that such
unconsciousness is usual in all cases where an action, and the design
which prompts it, have been repeated exceedingly often. If, again,
we are asked how we account for the regularity with which each step
is taken in its due order, we answer that this too is characteristic
of actions that are done habitually--they being very rarely misplaced
in respect of any part.

When I wrote "Life and Habit," I had arrived at the conclusion that
memory was the most essential characteristic of life, and went so far
as to say, "Life is that property of matter whereby it can remember--
matter which can remember is living." I should perhaps have written,
"Life is the being possessed of a memory--the life of a thing at any
moment is the memories which at that moment it retains"; and I would
modify the words that immediately follow, namely, "Matter which
cannot remember is dead"; for they imply that there is such a thing
as matter which cannot remember anything at all, and this on fuller
consideration I do not believe to be the case; I can conceive of no
matter which is not able to remember a little, and which is not
living in respect of what it can remember. I do not see how action
of any kind is conceivable without the supposition that every atom
retains a memory of certain antecedents. I cannot, however, at this
point, enter upon the reasons which have compelled me to this
conclusion. Whether these would be deemed sufficient or no, at any
rate we cannot believe that a system of self-reproducing associations
should develop from the simplicity of the amoeba to the complexity of
the human body without the presence of that memory which can alone
account at once for the resemblances and the differences between
successive generations, for the arising and the accumulation of
divergences--for the tendency to differ and the tendency not to

differ.

At parting, therefore, I would recommend the reader to see every atom in the universe as living and able to feel and to remember, but in a humble way. He must have life eternal, as well as matter eternal; and the life and the matter must be joined together inseparably as body and soul to one another. Thus he will see God everywhere, not as those who repeat phrases conventionally, but as people who would have their words taken according to their most natural and legitimate meaning; and he will feel that the main difference between him and many of those who oppose him lies in the fact that whereas both he and they use the same language, his opponents only half mean what they say, while he means it entirely.

The attempt to get a higher form of a life from a lower one is in accordance with our observation and experience. It is therefore proper to be believed. The attempt to get it from that which has absolutely no life is like trying to get something out of nothing. The millionth part of a farthing put out to interest at ten per cent, will in five hundred years become over a million pounds, and so long as we have any millionth of a millionth of the farthing to start with, our getting as many million pounds as we have a fancy for is only a question of time, but without the initial millionth of a millionth of a millionth part, we shall get no increment whatever. A little leaven will leaven the whole lump, but there must be SOME leaven.

I will here quote two passages from an article already quoted from on page 55 of this book. They run:-

"We are growing conscious that our earnest and most determined efforts to make motion produce sensation and volition have proved a

failure, and now we want to rest a little in the opposite, much less
laborious conjecture, and allow any kind of motion to start into
existence, or at least to receive its specific direction from
psychical sources; sensation and volition being for the purpose
quietly insinuated into the constitution of the ultimately moving
particles." {177a}

And:-

"In this light it can remain no longer surprising that we actually
find motility and sensibility so intimately interblended in nature."
{177b}

We should endeavour to see the so-called inorganic as living, in
respect of the qualities it has in common with the organic, rather
than the organic as non-living in respect of the qualities it has in
common with the inorganic. True, it would be hard to place one's
self on the same moral platform as a stone, but this is not
necessary; it is enough that we should feel the stone to have a moral
platform of its own, though that platform embraces little more than a
profound respect for the laws of gravitation, chemical affinity, &c.
As for the difficulty of conceiving a body as living that has not got
a reproductive system--we should remember that neuter insects are
living but are believed to have no reproductive system. Again, we
should bear in mind that mere assimilation involves all the
essentials of reproduction, and that both air and water possess this
power in a very high degree. The essence of a reproductive system,
then, is found low down in the scheme of nature.

At present our leading men of science are in this difficulty; on the

one hand their experiments and their theories alike teach them that
spontaneous generation ought not to be accepted; on the other, they
must have an origin for the life of the living forms, which, by their
own theory, have been evolved, and they can at present get this
origin in no other way than by the Deus ex machina method, which they
reject as unproved, or a spontaneous generation of living from non-
living matter, which is no less foreign to their experience. As a
general rule, they prefer the latter alternative. So Professor
Tyndall, in his celebrated article (Nineteenth Century, November
1878), wrote:-

"It is generally conceded (and seems to be a necessary inference from
the lessons of science) that SPONTANEOUS GENERATION MUST AT ONE
TIME
HAVE TAKEN PLACE" (italics mine).

No inference can well be more unnecessary or unscientific. I suppose
spontaneous generation ceases to be objectionable if it was "only a
very little one," and came off a long time ago in a foreign country.
The proper inference is, that there is a low kind of livingness in
every atom of matter. Life eternal is as inevitable a conclusion as
matter eternal.

It should not be doubted that wherever there is vibration or motion
there is life and memory, and that there is vibration and motion at
all times in all things.

The reader who takes the above position will find that he can explain
the entry of what he calls death among what he calls the living,
whereas he could by no means introduce life into his system if he
started without it. Death is deducible; life is not deducible.
Death is a change of memories; it is not the destruction of all
memory. It is as the liquidation of one company, each member of

which will presently join a new one, and retain a trifle even of the old cancelled memory, by way of greater aptitude for working in concert with other molecules. This is why animals feed on grass and on each other, and cannot proselytise or convert the rude ground before it has been tutored in the first principles of the higher kinds of association.

Again, I would recommend the reader to beware of believing anything in this book unless he either likes it, or feels angry at being told it. If required belief in this or that makes a man angry, I suppose he should, as a general rule, swallow it whole then and there upon the spot, otherwise he may take it or leave it as he likes. I have not gone far for my facts, nor yet far from them; all on which I rest are as open to the reader as to me. If I have sometimes used hard terms, the probability is that I have not understood them, but have done so by a slip, as one who has caught a bad habit from the company he has been lately keeping. They should be skipped.

Do not let him be too much cast down by the bad language with which professional scientists obscure the issue, nor by their seeming to make it their business to fog us under the pretext of removing our difficulties. It is not the ratcatcher's interest to catch all the rats; and, as Handel observed so sensibly, "Every professional gentleman must do his best for to live." The art of some of our philosophers, however, is sufficiently transparent, and consists too often in saying "organism which must be classified among fishes," instead of "fish," {179a} and then proclaiming that they have "an ineradicable tendency to try to make things clear." {179b}

If another example is required, here is the following from an article than which I have seen few with which I more completely agree, or which have given me greater pleasure. If our men of science would take to writing in this way, we should be glad enough to follow them.

The passage I refer to runs thus:-

"Professor Huxley speaks of a 'verbal fog by which the question at
issue may be hidden'; is there no verbal fog in the statement that
THE AETIOLOGY OF CRAYFISHES RESOLVES ITSELF INTO A GRADUAL
EVOLUTION
IN THE COURSE OF THE MESOSOIC AND SUBSEQUENT EPOCHS OF THE
WORLD'S
HISTORY OF THESE ANIMALS FROM A PRIMITIVE ASTACOMORPHOUS
FORM? Would
it be fog or light that would envelop the history of man if we said
that the existence of man was explained by the hypothesis of his
gradual evolution from a primitive anthropomorphous form? I should
call this fog, not light." {180}

Especially let him mistrust those who are holding forth about
protoplasm, and maintaining that this is the only living substance.
Protoplasm may be, and perhaps is, the MOST living part of an
organism, as the most capable of retaining vibrations, but this is
the utmost that can be claimed for it.

Having mentioned protoplasm, I may ask the reader to note the
breakdown of that school of philosophy which divided the ego from the
non ego. The protoplasmists, on the one hand, are whittling away at
the ego, till they have reduced it to a little jelly in certain parts
of the body, and they will whittle away this too presently, if they
go on as they are doing now.

Others, again, are so unifying the ego and the non ego, that with
them there will soon be as little of the non ego left as there is of
the ego with their opponents. Both, however, are so far agreed as

that we know not where to draw the line between the two, and this renders nugatory any system which is founded upon a distinction between them.

The truth is, that all classification whatever, when we examine its raison d'etre closely, is found to be arbitrary--to depend on our sense of our own convenience, and not on any inherent distinction in the nature of the things themselves. Strictly speaking, there is only one thing and one action. The universe, or God, and the action of the universe as a whole.

Lastly, I may predict with some certainty that before long we shall find the original Darwinism of Dr. Erasmus Darwin (with an infusion of Professor Hering into the bargain) generally accepted instead of the neo-Darwinism of to-day, and that the variations whose accumulation results in species will be recognised as due to the wants and endeavours of the living forms in which they appear, instead of being ascribed to chance, or, in other words, to unknown causes, as by Mr. Charles Darwin's system. We shall have some idyllic young naturalist bringing up Dr. Erasmus Darwin's note on Trapa natans, {181a} and Lamarck's kindred passage on the descent of Ranunculus hederaceus from Ranunculus aquatilis {181b} as fresh discoveries, and be told, with much happy simplicity, that those animals and plants which have felt the need of such or such a structure have developed it, while those which have not wanted it have gone without it. Thus, it will be declared, every leaf we see around us, every structure of the minutest insect, will bear witness to the truth of the "great guess" of the greatest of naturalists concerning the memory of living matter.

I dare say the public will not object to this, and am very sure that none of the admirers of Mr. Charles Darwin or Mr. Wallace will protest against it; but it may be as well to point out that this was

not the view of the matter taken by Mr. Wallace in 1858 when he and
Mr. Darwin first came forward as preachers of natural selection. At
that time Mr. Wallace saw clearly enough the difference between the
theory of "natural selection" and that of Lamarck. He wrote:-

"The hypothesis of Lamarck--that progressive changes in species have
been produced by the attempts of animals to increase the development
of their own organs, and thus modify their structure and habits--has
been repeatedly and easily refuted by all writers on the subject of
varieties and species, . . . but the view here developed tenders such
an hypothesis quite unnecessary. . . . The powerful retractile
talons of the falcon and the cat tribes have not been produced or
increased by the volition of those animals, neither did the giraffe
acquire its long neck by desiring to reach the foliage of the more
lofty shrubs, and constantly stretching its neck for this purpose,
but because any varieties which occurred among its antitypes with a
longer neck than usual AT ONCE SECURED A FRESH RANGE OF PASTURE
OVER
THE SAME GROUND AS THEIR SHORTER-NECKED COMPANIONS, AND
ON THE FIRST
SCARCITY OF FOOD WERE THEREBY ENABLED TO OUTLIVE THEM" (italics in
original). {182a}

This is absolutely the neo-Darwinian doctrine, and a denial of the
mainly fortuitous character of the variations in animal and vegetable
forms cuts at its root. That Mr. Wallace, after years of reflection,
still adhered to this view, is proved by his heading a reprint of the
paragraph just quoted from {182b} with the words "Lamarck's
hypothesis very different from that now advanced"; nor do any of his
more recent works show that he has modified his opinion. It should

be noted that Mr. Wallace does not call his work "Contributions to the Theory of Evolution," but to that of "Natural Selection."

Mr. Darwin, with characteristic caution, only commits himself to saying that Mr. Wallace has arrived at ALMOST (italics mine) the same general conclusions as he, Mr. Darwin, has done; {182c} but he still, as in 1859, declares that it would be "a serious error to suppose that the greater number of instincts have been acquired by habit in one generation, and then transmitted by inheritance to succeeding generations," {183a} and he still comprehensively condemns the "well-known doctrine of inherited habit, as advanced by Lamarck." {183b}

As for the statement in the passage quoted from Mr. Wallace, to the effect that Lamarck's hypothesis "has been repeatedly and easily refuted by all writers on the subject of varieties and species," it is a very surprising one. I have searched Evolution literature in vain for any refutation of the Erasmus Darwinian system (for this is what Lamarck's hypothesis really is) which need make the defenders of that system at all uneasy. The best attempt at an answer to Erasmus Darwin that has yet been made is "Paley's Natural Theology," which was throughout obviously written to meet Buffon and the "Zoonomia." It is the manner of theologians to say that such and such an objection "has been refuted over and over again," without at the same time telling us when and where; it is to be regretted that Mr. Wallace has here taken a leaf out of the theologians' book. His statement is one which will not pass muster with those whom public opinion is sure in the end to follow.

Did Mr. Herbert Spencer, for example, "repeatedly and easily refute" Lamarck's hypothesis in his brilliant article in the Leader, March 20, 1852? On the contrary, that article is expressly directed against those "who cavalierly reject the hypothesis of Lamarck and his followers." This article was written six years before the words last quoted from Mr. Wallace; how absolutely, however, does the word "cavalierly" apply to them!

Does Isidore Geoffroy, again, bear Mr. Wallace's assertion out better? In 1859--that is to say, but a short time after Mr. Wallace had written--he wrote as follows:-

"Such was the language which Lamarck heard during his protracted old age, saddened alike by the weight of years and blindness; this was what people did not hesitate to utter over his grave yet barely closed, and what indeed they are still saying--commonly too without any knowledge of what Lamarck maintained, but merely repeating at secondhand bad caricatures of his teaching.

"When will the time come when we may see Lamarck's theory discussed-- and, I may as well at once say, refuted in some important points {184a}--with at any rate the respect due to one of the most illustrious masters of our science? And when will this theory, the hardihood of which has been greatly exaggerated, become freed from the interpretations and commentaries by the false light of which so many naturalists have formed their opinion concerning it? If its author is to be condemned, let it be, at any rate, not before he has been heard." {184b}

In 1873 M. Martin published his edition of Lamarck's "Philosophie Zoologique." He was still able to say, with, I believe, perfect truth, that Lamarck's theory has "never yet had the honour of being discussed seriously." {184c}

Professor Huxley in his article on Evolution is no less cavalier than Mr. Wallace. He writes:- {184d}

"Lamarck introduced the conception of the action of an animal on itself as a factor in producing modification."

[Lamarck did nothing of the kind. It was Buffon and Dr. Darwin who introduced this, but more especially Dr. Darwin.]

"But A LITTLE CONSIDERATION SHOWED" (italics mine) "that though Lamarck had seized what, as far as it goes, is a true cause of modification, it is a cause the actual effects of which are wholly inadequate to account for any considerable modification in animals, and which can have no influence whatever in the vegetable world, &c."

I should be very glad to come across some of the "little consideration" which will show this. I have searched for it far and wide, and have never been able to find it.

I think Professor Huxley has been exercising some of his ineradicable tendency to try to make things clear in the article on Evolution, already so often quoted from. We find him (p. 750) pooh-poohing Lamarck, yet on the next page he says, "How far 'natural selection' suffices for the production of species remains to be seen." And this when "natural selection" was already so nearly of age! Why, to those who know how to read between a philosopher's lines, the sentence comes to very nearly the same as a declaration that the writer has no great opinion of "natural selection." Professor Huxley continues, "Few can doubt that, if not the whole cause, it is a very important factor in that operation." A philosopher's words should be weighed carefully, and when Professor Huxley says "few can doubt," we must remember that he may be including himself among the few whom he considers to have the power of doubting on this matter. He does not say "few will," but "few can" doubt, as though it were only the enlightened who would have the power of doing so. Certainly "nature,"--for this is what "natural selection" comes to,--is rather an important factor in the operation, but we do not gain much by being told so. If, however, Professor Huxley neither believes in the origin of species, through sense of need on the part of animals themselves, nor yet in "natural selection," we should be glad to know what he does believe in.

The battle is one of greater importance than appears at first sight. It is a battle between teleology and non-teleology, between the purposiveness and the non-purposiveness of the organs in animal and vegetable bodies. According to Erasmus Darwin, Lamarck, and Paley, organs are purposive; according to Mr. Darwin and his followers, they are not purposive. But the main arguments against the system of Dr. Erasmus Darwin are arguments which, so far as they have any weight, tell against evolution generally. Now that these have been disposed of, and the prejudice against evolution has been overcome, it will be seen that there is nothing to be said against the system of Dr. Darwin and Lamarck which does not tell with far greater force against that of Mr. Charles Darwin and Mr. Wallace.

Notes:

{0a} This is the date on the title-page. The preface is dated October 15, 1886, and the first copy was issued in November of the same year. All the dates are taken from the Bibliography by Mr. H. Festing Jones prefixed to the "Extracts" in the New Quarterly Review (1909).

{0b} I.e. after p. 285: it bears no number of its own!

{0c} The distinction was merely implicit in his published writings, but has been printed since his death from his "Notebooks," New Quarterly Review, April, 1908. I had developed this thesis, without knowing of Butler's explicit anticipation in an article then in the press: "Mechanism and Life," Contemporary Review, May, 1908.

{0d} The term has recently been revived by Prof. Hubrecht and by myself (Contemporary Review, November 1908).

{0e} See Fortnightly Review, February 1908, and Contemporary Review, September and November 1909. Since these publications the hypnosis seems to have somewhat weakened.

{0f} A "hormone" is a chemical substance which, formed in one part of the body, alters the reactions of another part, normally for the good of the organism.

{0g} Mr. H. Festing Jones first directed my attention to these passages and their bearing on the Mutation Theory.

{0i} He says in a note, "This general type of reaction was described and illustrated in a different connection by Pfluger in 'Pfluger's Archiv. f.d. ges. Physiologie,' Bd. XV." The essay bears the significant title "Die teleologische Mechanik der lebendigen Natur," and is a very remarkable one, as coming from an official physiologist in 1877, when the chemico-physical school was nearly at its zenith.

{0j} "Contributions to the Study of the Lower Animals" (1904), "Modifiability in Behaviour" and "Method of Regulability in Behaviour and in other Fields," in Journ. Experimental Zoology, vol. ii.

(1905).

{0h} See "The Hereditary Transmission of Acquired Characters" in Contemporary Review, September and November 1908, in which references are given to earlier statements.

{0k} Semon's technical terms are exclusively taken from the Greek, but as experience tells that plain men in England have a special dread of suchlike, I have substituted "imprint" for "engram," "outcome" for "ecphoria"; for the latter term I had thought of "efference," "manifestation," etc., but decided on what looked more homely, and at the same time was quite distinctive enough to avoid that confusion which Semon has dodged with his Graecisms.

{0l} "Between the 'me' of to-day and the 'me' of yesterday lie night and sleep, abysses of unconsciousness; nor is there any bridge but memory with which to span them."--Unconscious Memory, p. 71.

{0m} Preface by Mr. Charles Darwin to "Erasmus Darwin." The Museum has copies of a Kosmos that was published 1857-60 and then discontinued; but this is clearly not the Kosmos referred to by Mr. Darwin, which began to appear in 1878.

{0n} Preface to "Erasmus Darwin."

{2} May 1880.

{3} Kosmos, February 1879, Leipsic.

{4} Origin of Species, ed. i., p. 459.

{8a} Origin of Species, ed. i., p. 1.

{8b} Kosmos, February 1879, p. 397.

{8c} Erasmus Darwin, by Ernest Krause, pp. 132, 133.

{9a} Origin of Species, ed. i., p. 242.

{9b} Ibid., p. 427.

{10a} Nineteenth Century, November 1878; Evolution, Old and New, pp. 360. 361.

{10b} Encyclopaedia Britannica, ed. ix., art. "Evolution," p. 748.

{11} Ibid.

{17} Encycl. Brit., ed. ix., art. "Evolution," p. 750.

{23a} Origin of Species, 6th ed., 1876, p. 206.

{23b} Ibid., p. 233.

{24a} Origin of Species, 6th ed., p. 171, 1876.

{24b} Pp. 258-260.

{26} Zoonomia, vol. i. p. 484; Evolution, Old and New, p. 214.

{27} "Erasmus Darwin," by Ernest Krause, p. 211, London, 1879.

{28a} See "Evolution, Old and New," p. 91, and Buffon, tom. iv. p. 383, ed. 1753.

{28b} Evolution, Old and New, p. 104.

{29a} Encycl. Brit., 9th ed., art. "Evolution," p. 748.

{29b} Palingenesie Philosophique, part x. chap. ii. (quoted from Professor Huxley's article on "Evolution," Encycl. Brit., 9th ed., p. 745).

{31} The note began thus: "I have taken the date of the first publication of Lamarck from Isidore Geoffroy St. Hilaire's (Hist. Nat. Generale tom. ii. p. 405, 1859) excellent history of opinion upon this subject. In this work a full account is given of Buffon's fluctuating conclusions upon the same subject."--Origin of Species, 3d ed., 1861, p. xiv.

{33a} Life of Erasmus Darwin, pp. 84, 85.

{33b} See Life and Habit, p. 264 and pp. 276, 277.

{33c} See Evolution, Old and New, pp. 159-165.

{33d} Ibid., p. 122.

{34} See Evolution, Old and New, pp. 247, 248.

{35a} Vestiges of Creation, ed. 1860, "Proofs, Illustrations, &c.," p. lxiv.

{35b} The first announcement was in the Examiner, February 22, 1879.

{36} Saturday Review, May 31, 1879.

{37a} May 26, 1879.

{37b} May 31, 1879.

{37c} July 26, 1879.

{37d} July 1879.

{37e} July 1879.

{37f} July 29, 1879.

{37g} January 1880.

{39} How far Kosmos was "a well-known" journal, I cannot determine. It had just entered upon its second year.

{41} Evolution, Old and New, p. 120, line 5.

{43} Kosmos, February 1879, p. 397.

{44a} Kosmos, February 1879, p. 404.

{44b} Page 39 of this volume.

{50} See Appendix A.

{52} Since published as "God the Known and God the Unknown." Fifield, 1s. 6d. net. 1909.

{54a} "Contemplation of Nature," Engl. trans., Lond. 1776. Preface, p. xxxvi.

{54b} Ibid., p. xxxviii.

{55} Life and Habit, p. 97.

{56} "The Unity of the Organic Individual," by Edward Montgomery, Mind, October 1880, p. 466.

{58} Life and Habit, p. 237.

{59a} Discourse on the Study of Natural Philosophy. Lardner's Cab. Cyclo., vol. xcix. p. 24.

{59b} Young's Lectures on Natural Philosophy, ii. 627. See also Phil. Trans., 1801-2.

{63} The lecture is published by Karl Gerold's Sohn, Vienna.

{69} See quotation from Bonnet, p. 54 of this volume.

{70} Professor Hering is not clear here. Vibrations (if I understand his theory rightly) should not be set up by faint stimuli from within. Whence and what are these stimuli? The vibrations within are already existing, and it is they which are the stimuli to action. On having been once set up, they either continue in sufficient force to maintain action, or they die down, and become too weak to cause further action, and perhaps even to be perceived within the mind, until they receive an accession of vibration from without. The only "stimulus from within" that should be able to generate action is that which may follow when a vibration already established in the body runs into another similar vibration already so established. On this consciousness, and even action, might be supposed to follow without the presence of an external stimulus.

{71} This expression seems hardly applicable to the overtaking of an internal by an external vibration, but it is not inconsistent with it. Here, however, as frequently elsewhere, I doubt how far Professor Hering has fully realised his conception, beyond being, like myself, convinced that the phenomena of memory and of heredity have a common source.

{72} See quotation from Bonnet, p. 54 of this volume. By "preserving the memory of habitual actions" Professor Hering probably means, retains for a long while and repeats motion of a certain character when such motion has been once communicated to it.

{74a} It should not be "if the central nerve system were not able to reproduce whole series of vibrations," but "if whole series of vibrations do not persist though unperceived," if Professor Hering intends what I suppose him to intend.

{74b} Memory was in full operation for so long a time before anything like what we call a nervous system can be detected, that Professor Hering must not be supposed to be intending to confine memory to a motor nerve system. His words do not even imply that he does, but it is as well to be on one's guard.

{77} It is from such passages as this, and those that follow on the next few pages, that I collect the impression of Professor Hering's meaning which I have endeavoured to convey in the preceding chapter.

{78} That is to say, "an infinitely small change in the kind of vibration communicated from the parent to the germ."

{79} It may be asked what is meant by responding. I may repeat that I understand Professor Hering to mean that there exists in the offspring certain vibrations, which are many of them too faint to upset equilibrium and thus generate action, until they receive an accession of force from without by the running into them of vibrations of similar characteristics to their own, which last vibrations have been set up by exterior objects. On this they become strong enough to generate that corporeal earthquake which we call action.

This may be true or not, but it is at any rate intelligible; whereas much that is written about "fraying channels" raises no definite ideas in the mind.

{80a} I interpret this, "We cannot wonder if often-repeated vibrations gather strength, and become at once more lasting and requiring less accession of vibration from without, in order to become strong enough to generate action."

{80b} "Characteristics" must, I imagine, according to Professor Hering, resolve themselves ultimately into "vibrations," for the characteristics depend upon the character of the vibrations.

{81} Professor Hartog tells me that this probably refers to Fritz Muller's formulation of the "recapitulation process" in "Facts for Darwin," English edition (1869), p. 114.--R.A.S.

{82} This is the passage which makes me suppose Professor Hering to mean that vibrations from exterior objects run into vibrations already existing within the living body, and that the accession to power thus derived is his key to an explanation of the physical basis of action.

{84} I interpret this: "There are fewer vibrations persistent within the bodies of the lower animals; those that there are, therefore, are stronger and more capable of generating action or upsetting the status in quo. Hence also they require less accession of vibration from without. Man is agitated by more and more varied vibrations; these, interfering, as to some extent they must, with one another, are weaker, and therefore require more accession from without before they can set the mechanical adjustments of the body in motion."

{89} I am obliged to Mr. Sully for this excellent translation of "Hellsehen."

{90a} Westminster Review, New Series, vol. xlix. p. 143.

{90b} Ibid., p. 145.

{90c} Ibid., p. 151.

{92a} "Instinct ist zweckmassiges Handeln ohne Bewusstsein des Zwecks."--Philosophy of the Unconscious, 3d ed., Berlin, 1871, p. 70.

{92b} "1. Eine blosse Folge der korperlichen Organisation.

"2. Ein von der Natur eingerichteter Gehirn-oder Geistesmechanismus.

"3. Eine Folge unbewusster Geistesthiitigkeit."--Philosophy of the Unconscious, 3d ed., p. 70.

{97} "Hiermit ist der Annahme das Urtheil gesprochen, welche die unbewusste Vorstellung des Zwecks in jedem einzelnen Falle vorwiegt; denn wollte man nun noch die Vorstellung des Geistesmechanismus festhalten so musste fur jede Variation und Modification des Instincts, nach den ausseren Umstanden, eine besondere constante Vorrichtung . . . eingefugt sein."--Philosophy of the Unconscious 3d ed., p. 74.

{99} "Indessen glaube ich, dass die angefuhrten Beispiele zur Genuge

beweisen, dass es auch viele Falle giebt, wo ohne jede Complication
mit der bewussten Ueberlegung die gewohnliche und aussergewohnliche
Handlung aus derselben Quelle stammen, dass sie entweder beide
wirklicher Instinct, oder beide Resultate bewusster Ueberlegung
sind."--Philosophy of the Unconscious, 3d ed., p. 76.

{100} "Dagegen haben wir nunmehr unseren Blick noch einmal scharfer
auf den Begriff eines psychischen Mechanismus zu richten, und da
zeigt sich, dass derselbe, abgesehen davon, wie viel er erklart, so
dunke list, dass man sich kaum etwas dabei denken kann."--Philosophy
of the Unconscious, 3d ed., p. 76.

{101} "Das Endglied tritt als bewusster Wille zu irgend einer
Handlung auf; beide sind aber ganz ungleichartig und haben mit der
gewohnlichen Motivation nichts zu thun, welche ausschliesslich darin
besteht, dass die Vorstellung einer Lust oder einer Unlust das
Begehren erzeugr, erstere zu erlangen, letztere sich fern zu
halten."--Ibid., p. 76.

{102a} "Diese causale Verbindung fallt erfahrungsmassig, wie wir von
unsern menschlichen Instincten wissen, nicht in's Bewussisein;
folglich kann dieselbe, wenn sie ein Mechanismus sein soll, nur
entweder ein nicht in's Bewusstsein fallende mechanische Leitung und
Umwandlung der Schwingungen des vorgestellten Motivs in die
Schwingungen der gewollten Handlung im Gehirn, oder ein unbewusster
geistiger Mechanismus sein."--Philosophy of the Unconscious 3d ed.,
p. 77.

{102b} "Man hat sich also zwischen dem bewussten Motiv, und dem
Willen zur Insticthandlung eine causale Verbindung durch unbewusstes
Vorstellen und Wollen zu denken, und ich weiss nicht, wie diese
Verbindung einfacher gedacht werden konnte, als durch den
vorgestellten und gewollten Zweck. Damit sind wir aber bei dem allen

Geistern eigenthumlichen und immanenten Mechanismus der Logik
angelangt, und haben die unbewusster Zweckvorstellung bei jeder
einzelnen Instincthandlung als unentbehrliches Glied gefunden;
hiermit hat also der Begrift des todten, ausserlich pradestinirten
Geistesmechanismus sich selbst aufgehoben und in das immanente
Geistesleben der Logik umgewandelt, und wir sind bei der letzten
Moglichkeit angekommen, welche fur die Auffassung eines wirklichen
Instincts ubrig bleibt: der Instinct ist bewusstes Wollen des
Mittels zu einem unbewusst gewollten Zweck."--Philosophy of the
Unconscious, 3d ed., p. 78.

{105a} "Also der Instinct ohne Hulfsmechanismus die Ursache der
Entstehung des Hulfsmechanismus ist."--Philosophy of the Unconscious,
3d ed., p. 79.

{105b} "Dass auch der fertige Hulfsmechanismus das Unbewusste nicht
etwa zu dieser bestimmten Instincthandlung necessirt, sondern blosse
pradisponirt."--Philosophy of the Unconscious, 3d ed., p. 79.

{105c} "Giebt es einen wirklichen Instinct, oder sind die
sogenannten Instincthandlungen nur Resultate bewusster Ueberlegung?"-
-Philosophy of the Unconscious, 3d ed., p. 79.

{111} "Dieser Beweis ist dadurch zu fuhren; erstens dass die
betreffenden Thatsachen in; der Zukunft liegen, und dem Verstande die
Anhaltepunkte fehlen, um ihr zukunftiges Eintreten aus den
gegenwartigen Verhaltnissen zu erschliessen; zweitens, dass die
betreffenden Thatsachen augenscheinlich der sinnlichen Wahrnehmung
verschlossen liegen, weil nur die Erfahrung fruherer Falle uber sie
belehren kann, und diese laut der Beobachtung ausgeschlossen ist. Es
wurde fur unsere Interessen keinen Unterschied machen, wenn, was ich
wahrscheinlich halte, bei fortschreitender physiologischer
Erkenntniss alle jetzt fur den ersten Fall anzufuhrenden Beispiele

sich als solche des zweiten Falls ausweisen sollten, wie dies
unleugbar bei vielen fruher gebrauchten Beispielen schon geschehen
ist; denn ein apriorisches Wissen ohne jeden sinnlichen Anstoss ist
wohl kaum wunderbarer zu nennen, als ein Wissen, welches zwar BEI
GELEGENHEIT gewisser sinnlicher Wahrnehmung zu Tage tritt, aber mit
diesen nur durch eine solche Kette von Schlussen und angewandten
Kenntnissen in Verbindung stehend gedacht werden konnte, dass deren
Moglichkeit bei dem Zustande der Fahigkeiten und Bildung der
betreffenden Thiere entschieden geleugnet werden muss."--Philosophy
of the Unconscious, 3d ed., p. 85.

{113} "Man hat dieselbe jederzeit anerkannt und mit den Worten
Vorgefuhl oder Ahnung bezeichnet; indess beziehen sich diese Worte
einerseits nur auf zukunftiges, nicht auf gegenwartiges, raumlich
getrennte Unwahmehrnbares, anderseits bezeichnen sie nur die leise,
dumpfe, unbestimmte Resonanz des Bewusstseins mit dem unfehlbar
bestimmten Zustande der unbewussten Erkenntniss. Daher das Wort
Vorgefuhl in Rucksicht auf die Dumpfheit und Unbestimmtheit, wahrend
doch leicht zu sehen ist, dass das von allen, auch den unbewussten
Vorstellungen entblosste Gefuhl fur das Resultat gar keinen Einfluss
haben kann, sondern nur eine Vorstellung, weil diese allein
Erkenntniss enthalt. Die in Bewusstsein mitklingende Ahnung kann
allerdings unter Umstanden ziemlich deutlich sein, so dass sie sich
beim Menschen in Gedanken und Wort fixiren lasst; doch ist dies auch
im Menschen erfahrungsmassig bei den eigenthumlichen Instincten nicht
der Fall, vielmehr ist bei diesen die Resonanz der unbewussten
Erkenntniss im Bewusstsein meistens so schwach, dass sie sich
wirklich nur in begleitenden Gefuhlen oder der Stimmung aussert, dass
sie einen unendlich kleinen Bruchtheil des Gemeingefuhls bildet."--
Philosophy of the Unconscious, 3d ed., p. 86.

{115a} "In der Bestimmung des Willens durch einen im Unbewussten
liegenden Process . . . fur welchen sich dieser Character der

zweifellosen Selbstgewissheit in allen folgenden Untersuchungen bewahren wird."--Philosophy of the Unconscious, p. 87.

{115b} "Sondern als unmittelbarer Besitz vorgefunden wird."--Philosophy of the Unconscious, p. 87.

{115c} "Hellsehen."

{119a} "Das Hellsehon des Unbewussten hat sie den rechten Weg ahnen lassen."--Philosophy of the Unconscious, p. 90, 3d ed., 1871.

{119b} "Man wird doch wahrlich nicht den Thieren zumuthen wollen, durch meteorologische Schlusse das Wetter auf Monate im Voraus zu berechnen, ja sogar Ueberschwemmungen vorauszusehen. Vielmehr ist eine solche Gefuhlswahrnehmung gegenwartiger atmospharischer Einflusse nichts weiter als die sinnliche Wahrnehmung, welche als Motiv wirkt, und ein Motiv muss ja doch immer vorhanden sein, wenn ein Instinct functioniren soll. Es bleibt also trotzdem bestehen dass das Voraussehen der Witterung ein unbewusstes Hellsehen ist, von dem der Storch, der vier Wochen fruher nach Suden aufbricht, so wenig etwas weiss, als der Hirsch, der sich vor einem kalten Winter einen dickeren Pelz als gewohnlich wachsen lasst. Die Thiere haben eben einerseits das gegenwartige Witterungsgefuhl im Bewusstsein, daraus folgt andererseits ihr Handeln gerade so, als ob sie die Vorstellung der zukunftigen Witterung hatten; im Bewusstsein haben sie dieselbe aber nicht, also bietet sich als einzig naturliches Mittelglied die unbewusste Vorstellung, die nun aber immer ein Hellsehen ist, weil sie etwas enthalt, was dem Thier weder dutch sinnliche Wahrnehmung direct gegeben ist, noch durch seine Verstandesmittel aus der Wahrnehmung geschlossen werden kann."--Philosophy of the Unconscious, p. 91, 3d ed., 1871.

{124} "Meistentheils tritt aber hier der hoheren Bewusstseinstufe

der Menschen entsprechend eine starkete Resonanz des Bewusstseins mit
dem bewussten Hellsehen hervor, die sich also mehr odor minder
deutliche Ahnung darstellt. Ausserdem entspricht es der grosseren
Selbststandigkeit des menschlichen Intellects, dass diese Ahnung
nicht ausschliesslich Behufs der unmittelbaren Ausfuhrung einer
Handlung eintritt, sondern bisweilen auch unabangig von der Bedingung
einer momentan zu leistenden That als blosse Vorstellung ohne
bewussten Willen sich zeigte, wenn nur die Bedingung erfullt ist,
dass der Gegenstand dieses Ahnens den Willen des Ahnenden im
Allgemeinen in hohem Grade interessirt."--Philosophy of the
Unconscious, 3d ed., p. 94.

{126} "Haufig sind die Ahnungen, in denen das Hellsehen des
Unbewussten sich dem Bewusstsein offenbart, dunkel, unverstandlich
und symbolisch, weil sie im Gehirn sinnliche Form annehmen mussen,
wahrend die unbewusste Vorstellung an der Form der Sinnlichkeit kein
Theil haben kann."--Philosophy of the Unconscious, 3d ed., p. 96.

{128} "Ebenso weil es diese Reihe nur in gesteigerter
Bewusstseinresonanz fortsetzt, stutzt es jene Aussagen der
Instincthandlungen uher ihr eigenes Wesen ebenso sehr," &c.--
Philosophy of the Unconscious, 3d ed., p. 97.

{129} "Wir werden trotzdem diese gomeinsame Wirkung eines
Masseninstincts in der Entstehung der Sprache und den grossen
politischen und socialen Bewegungen in der Woltgeschichte deutlich
wieder erkennen; hier handelt es sich um moglichst einfache und
deutliche Beispiele, und darum greifen wir zu niederen Thieren, wo
die Mittel der Gedankenmittheilung bei fehlender Stimme, Mimik und
Physiognomie so unvollkommen sind, dass die Uebereinstimmung und das
Ineinandergreifen der einzelnen Leistungen in den Hauptsachen
unmoglich der bewussten Verstandigung durch Sprache zugeschrieben
werden darf."--Philosophy of the Unconscious, 3d ed., p. 98.

{131a} "Und wie durch Instinct dot Plan des ganzen Stocks in unbewusstem Hellsehen jeder einzelnen Biene einwohnt."--Philosophy of the Unconscious, 3d ed., p. 99.

{131b} "Indem jedes Individuum den Plan des Ganzen und Sammtliche gegenwartig zu ergreifende Mittel im unbewussten Hellsehen hat, wovon aber nut das Eine, was ihm zu thun obliegt, in sein Bewusstsein fallt."--Philosophy of the Unconscious, 3d ed., p. 99.

{132} "Der Instinct ist nicht Resultat bewusster Ueberlegung, nicht Folge der korperlichen Organisation, nicht blosses Resultat eines in der Organisation des Gehirns gelegenen Mechanismus, nicht Wirkung eines dem Geiste von aussen angeklebten todten, seinem innersten Wesen fremden Mechanismus, sondern selbsteigene Leistung des Individuum aus seinem innersten Wesen und Character entspringend."--Philosophy of the Unconscious, 3d ed., p. 100.

{133} "Haufig ist die Kenntniss des Zwecks der bewussten Erkenntniss durch sinnliche Wahrnehmung gar nicht zuganglich; dann documentirt sich die Eigenthumlichkeit des Unbewussten im Hellsehen, von welchem das Bewusstsein theils nar eine verschwindend dumpfe, theils auch namentlich beim Menschen mehr oder minder deutliche Resonanz als Ahnung versputt."--Philosophy of the Unconscious, 3d ed., p. 100.

{135} "Und eine so damonische Gewalt sollte durch etwas ausgeubt werden konnon, was als ein dem inneren Wesen fremder Mechanismus dem Geiste aufgepfropft ist, oder gar durch eine bewusste Ueberlegung, welche doch stets nur im kahlen Egoismus stecken bleibt," &c.--Philosophy of the Unconscious, 3d ed., p. 101.

{139a} Page 100 of this vol.

{139b} Pp. 106, 107 of this vol.

{140} Page 100 of this vol.

{141} Page 99 of this vol.

{144a} See page 115 of this volume.

{144b} Page 104 of this vol.

{146} The Spirit of Nature. J. A. Churchill & Co., 1880, p. 39.

{149} I have put these words into the mouth of my supposed objector, and shall put others like them, because they are characteristic; but nothing can become so well known as to escape being an inference.

{153} Erewhon, chap. xxiii.

{160} It must be remembered that this passage is put as if in the mouth of an objector.

{177a} "The Unity of the Organic Individual," by Edward Montgomery. Mind, October 1880, p. 477.

{177b} Ibid., p. 483.

{179a} Professor Huxley, Encycl. Brit., 9th ed., art. Evolution, p. 750.

{179b} "Hume," by Professor Huxley, p. 45.

{180} "The Philosophy of Crayfishes," by the Right Rev. the Lord Bishop of Carlisle. Nineteenth Century for October 1880, p. 636.

{181a} Les Amours des Plantes, p. 360. Paris, 1800.

{181b} Philosophie Zoologique, tom. i. p. 231. Ed. M. Martin. Paris, 1873.

{182a} Journal of the Proceedings of the Linnean Society. Williams & Norgate, 1858, p. 61.

{182b} Contributions to the Theory of Natural Selection, 2d ed., 1871, p. 41.

{182c} Origin of Species, p. 1, ed. 1872.

{183a} Origin of Species, 6th ed., p. 206. I ought in fairness to Mr. Darwin to say that he does not hold the error to be quite as serious as he once did. It is now "a serious error" only; in 1859 it was "the most serious error."--Origin of Species, 1st ed., p. 209.

{183b} Origin of Species, 1st ed., p. 242; 6th ed., p. 233.

{184a} I never could find what these particular points were.

{184b} Isidore Geoffroy, Hist. Nat. Gen., tom. ii. p. 407, 1859.

{184c} M. Martin's edition of the "Philosophie Zoologique" (Paris, 1873), Introduction, p. vi.

{184d} Encyclopaedia Britannica, 9th ed., p. 750.

The Codes Of Hammurabi And Moses
W. W. Davies

QTY

The discovery of the Hammurabi Code is one of the greatest achievements of archaeology, and is of paramount interest, not only to the student of the Bible, but also to all those interested in ancient history...

Religion   **ISBN:** *1-59462-338-4*   **Pages:132**
MSRP $12.95

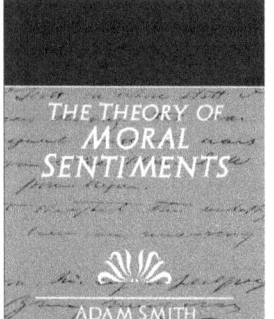

The Theory of Moral Sentiments
Adam Smith

QTY

This work from 1749. contains original theories of conscience amd moral judgment and it is the foundation for systemof morals.

Philosophy ISBN: *1-59462-777-0*   **Pages:536**
MSRP $19.95

Jessica's First Prayer
Hesba Stretton

QTY

In a screened and secluded corner of one of the many railway-bridges which span the streets of London there could be seen a few years ago, from five o'clock every morning until half past eight, a tidily set-out coffee-stall, consisting of a trestle and board, upon which stood two large tin cans, with a small fire of charcoal burning under each so as to keep the coffee boiling during the early hours of the morning when the work-people were thronging into the city on their way to their daily toil...

Childrens   ISBN: *1-59462-373-2*

Pages:84
MSRP $9.95

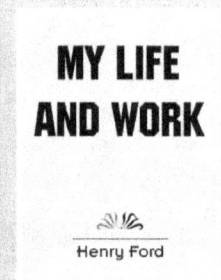

My Life and Work
Henry Ford

QTY

Henry Ford revolutionized the world with his implementation of mass production for the Model T automobile. Gain valuable business insight into his life and work with his own auto-biography... "We have only started on our development of our country we have not as yet, with all our talk of wonderful progress, done more than scratch the surface. The progress has been wonderful enough but..."

Pages:300

Biographies/   ISBN: *1-59462-198-5*   *MSRP $21.95*

www.bookjungle.com *email: sales@bookjungle.com fax: 630-214-0564 mail: Book Jungle PO Box 2226 Champaign, IL 61825*

The Art of Cross-Examination
Francis Wellman

QTY

I presume it is the experience of every author, after his first book is published upon an important subject, to be almost overwhelmed with a wealth of ideas and illustrations which could readily have been included in his book, and which to his own mind, at least, seem to make a second edition inevitable. Such certainly was the case with me; and when the first edition had reached its sixth impression in five months, I rejoiced to learn that it seemed to my publishers that the book had met with a sufficiently favorable reception to justify a second and considerably enlarged edition. ..

Reference   ISBN: *1-59462-647-2*

Pages:412

MSRP $19.95

On the Duty of Civil Disobedience
Henry David Thoreau

QTY

Thoreau wrote his famous essay, On the Duty of Civil Disobedience, as a protest against an unjust but popular war and the immoral but popular institution of slave-owning. He did more than write—he declined to pay his taxes, and was hauled off to gaol in consequence. Who can say how much this refusal of his hastened the end of the war and of slavery ?

Law   ISBN: *1-59462-747-9*

Pages:48

MSRP $7.45

Dream Psychology Psychoanalysis for Beginners
Sigmund Freud

QTY

Sigmund Freud, born Sigismund Schlomo Freud (May 6, 1856 - September 23, 1939), was a Jewish-Austrian neurologist and psychiatrist who co-founded the psychoanalytic school of psychology. Freud is best known for his theories of the unconscious mind, especially involving the mechanism of repression; his redefinition of sexual desire as mobile and directed towards a wide variety of objects; and his therapeutic techniques, especially his understanding of transference in the therapeutic relationship and the presumed value of dreams as sources of insight into unconscious desires.

Psychology   ISBN: *1-59462-905-6*

Pages:196

MSRP $15.45

The Miracle of Right Thought
Orison Swett Marden

QTY

Believe with all of your heart that you will do what you were made to do. When the mind has once formed the habit of holding cheerful, happy, prosperous pictures, it will not be easy to form the opposite habit. It does not matter how improbable or how far away this realization may see, or how dark the prospects may be, if we visualize them as best we can, as vividly as possible, hold tenaciously to them and vigorously struggle to attain them, they will gradually become actualized, realized in the life. But a desire, a longing without endeavor, a yearning abandoned or held indifferently will vanish without realization.

Self Help   ISBN: *1-59462-644-8*

Pages:360

MSRP $25.45

QTY

The Rosicrucian Cosmo-Conception Mystic Christianity by *Max Heindel* ISBN: *1-59462-188-8* **$38.95**
The Rosicrucian Cosmo-conception is not dogmatic, neither does it appeal to any other authority than the reason of the student. It is: not controversial, but is: sent forth in the, hope that it may help to clear... *New Age/Religion Pages 646*

Abandonment To Divine Providence by *Jean-Pierre de Caussade* ISBN: *1-59462-228-0* **$25.95**
"The Rev. Jean Pierre de Caussade was one of the most remarkable spiritual writers of the Society of Jesus in France in the 18th Century. His death took place at Toulouse in 1751. His works have gone through many editions and have been republished... *Inspirational/Religion Pages 400*

Mental Chemistry by *Charles Haanel* ISBN: *1-59462-192-6* **$23.95**
Mental Chemistry allows the change of material conditions by combining and appropriately utilizing the power of the mind. Much like applied chemistry creates something new and unique out of careful combinations of chemicals the mastery of mental chemistry... *New Age Pages 354*

The Letters of Robert Browning and Elizabeth Barret Barrett 1845-1846 vol II ISBN: *1-59462-193-4* **$35.95**
by *Robert Browning* and *Elizabeth Barrett* *Biographies Pages 596*

Gleanings In Genesis (volume I) by *Arthur W. Pink* ISBN: *1-59462-130-6* **$27.45**
Appropriately has Genesis been termed "the seed plot of the Bible" for in it we have, in germ form, almost all of the great doctrines which are afterwards fully developed in the books of Scripture which follow... *Religion/Inspirational Pages 420*

The Master Key by *L. W. de Laurence* ISBN: *1-59462-001-6* **$30.95**
In no branch of human knowledge has there been a more lively increase of the spirit of research during the past few years than in the study of Psychology, Concentration and Mental Discipline. The requests for authentic lessons in Thought Control, Mental Discipline and... *New Age/Business Pages 422*

The Lesser Key Of Solomon Goetia by *L. W. de Laurence* ISBN: *1-59462-092-X* **$9.95**
This translation of the first book of the "Lernegton" which is now for the first time made accessible to students of Talismanic Magic was done, after careful collation and edition, from numerous Ancient Manuscripts in Hebrew, Latin, and French... *New Age/Occult Pages 92*

Rubaiyat Of Omar Khayyam by *Edward Fitzgerald* ISBN:*1-59462-332-5* **$13.95**
Edward Fitzgerald, whom the world has already learned, in spite of his own efforts to remain within the shadow of anonymity, to look upon as one of the rarest poets of the century, was born at Bredfield, in Suffolk, on the 31st of March, 1809. He was the third son of John Purcell... *Music Pages 172*

Ancient Law by *Henry Maine* ISBN: *1-59462-128-4* **$29.95**
The chief object of the following pages is to indicate some of the earliest ideas of mankind, as they are reflected in Ancient Law, and to point out the relation of those ideas to modern thought. *Religiom/History Pages 452*

Far-Away Stories by *William J. Locke* ISBN: *1-59462-129-2* **$19.45**
"Good wine needs no bush, but a collection of mixed vintages does. And this book is just such a collection. Some of the stories I do not want to remain buried for ever in the museum files of dead magazine-numbers an author's not unpardonable vanity..." *Fiction Pages 272*

Life of David Crockett by *David Crockett* ISBN: *1-59462-250-7* **$27.45**
"Colonel David Crockett was one of the most remarkable men of the times in which he lived. Born in humble life, but gifted with a strong will, an indomitable courage, and unremitting perseverance... *Biographies/New Age Pages 424*

Lip-Reading by *Edward Nitchie* ISBN: *1-59462-206-X* **$25.95**
Edward B. Nitchie, founder of the New York School for the Hard of Hearing, now the Nitchie School of Lip-Reading, Inc, wrote "LIP-READING Principles and Practice". The development and perfecting of this meritorious work on lip-reading was an undertaking... *How-to Pages 400*

A Handbook of Suggestive Therapeutics, Applied Hypnotism, Psychic Science ISBN: *1-59462-214-0* **$24.95**
by *Henry Munro* *Health/New Age/Health/Self-help Pages 376*

A Doll's House: and Two Other Plays by *Henrik Ibsen* ISBN: *1-59462-112-8* **$19.95**
Henrik Ibsen created this classic when in revolutionary 1848 Rome. Introducing some striking concepts in playwriting for the realist genre, this play has been studied the world over. *Fiction/Classics/Plays 308*

The Light of Asia by *sir Edwin Arnold* ISBN: *1-59462-204-3* **$13.95**
In this poetic masterpiece, Edwin Arnold describes the life and teachings of Buddha. The man who was to become known as Buddha to the world was born as Prince Gautama of India but he rejected the worldly riches and abandoned the reigns of power when... *Religion/History/Biographies Pages 170*

The Complete Works of Guy de Maupassant by *Guy de Maupassant* ISBN: *1-59462-157-8* **$16.95**
"For days and days, nights and nights, I had dreamed of that first kiss which was to consecrate our engagement, and I knew not on what spot I should put my lips..." *Fiction/Classics Pages 240*

The Art of Cross-Examination by *Francis L. Wellman* ISBN: *1-59462-309-0* **$26.95**
Written by a renowned trial lawyer, Wellman imparts his experience and uses case studies to explain how to use psychology to extract desired information through questioning. *How-to/Science/Reference Pages 408*

Answered or Unanswered? by *Louisa Vaughan* ISBN: *1-59462-248-5* **$10.95**
Miracles of Faith in China *Religion Pages 112*

The Edinburgh Lectures on Mental Science (1909) by *Thomas* ISBN: *1-59462-008-3* **$11.95**
This book contains the substance of a course of lectures recently given by the writer in the Queen Street Hall, Edinburgh. Its purpose is to indicate the Natural Principles governing the relation between Mental Action and Material Conditions... *New Age/Psychology Pages 148*

Ayesha by *H. Rider Haggard* ISBN: *1-59462-301-5* **$24.95**
Verily and indeed it is the unexpected that happens! Probably if there was one person upon the earth from whom the Editor of this, and of a certain previous history, did not expect to hear again... *Classics Pages 380*

Ayala's Angel by *Anthony Trollope* ISBN: *1-59462-352-X* **$29.95**
The two girls were both pretty, but Lucy who was twenty-one who supposed to be simple and comparatively unattractive, whereas Ayala was credited, as her Bombwhat romantic name might show, with poetic charm and a taste for romance. Ayala when her father died was nineteen... *Fiction Pages 484*

The American Commonwealth by *James Bryce* ISBN: *1-59462-286-8* **$34.45**
An interpretation of American democratic political theory. It examines political mechanics and society from the perspective of Scotsman James Bryce *Politics Pages 572*

Stories of the Pilgrims by *Margaret P. Pumphrey* ISBN: *1-59462-116-0* **$17.95**
This book explores pilgrims religious oppression in England as well as their escape to Holland and eventual crossing to America on the Mayflower, and their early days in New England... *History Pages 268*

www.bookjungle.com *email: sales@bookjungle.com fax: 630-214-0564 mail: Book Jungle PO Box 2226 Champaign, IL 61825*

QTY

The Fasting Cure *by Sinclair Upton*   ISBN: *1-59462-222-1* **$13.95**
In the Cosmopolitan Magazine for May, 1910, and in the Contemporary Review (London) for April, 1910, I published an article dealing with my experiences in fasting. I have written a great many magazine articles, but never one which attracted so much attention... New Age/Self Help/Health Pages 164

Hebrew Astrology *by Sepharial*   ISBN: *1-59462-308-2* **$13.45**
In these days of advanced thinking it is a matter of common observation that we have left many of the old landmarks behind and that we are now pressing forward to greater heights and to a wider horizon than that which represented the mind-content of our progenitors... Astrology Pages 144

Thought Vibration or The Law of Attraction in the Thought World   ISBN: *1-59462-127-6* **$12.95**
by William Walker Atkinson   Psychology/Religion Pages 144

Optimism *by Helen Keller*   ISBN: *1-59462-108-X* **$15.95**
Helen Keller was blind, deaf, and mute since 19 months old, yet famously learned how to overcome these handicaps, communicate with the world, and spread her lectures promoting optimism. An inspiring read for everyone... Biographies/Inspirational Pages 84

Sara Crewe *by Frances Burnett*   ISBN: *1-59462-360-0*   **$9.45**
In the first place, Miss Minchin lived in London. Her home was a large, dull, tall one, in a large, dull square, where all the houses were alike, and all the sparrows were alike, and where all the door-knockers made the same heavy sound... Childrens/Classic Pages 88

The Autobiography of Benjamin Franklin *by Benjamin Franklin*   ISBN: *1-59462-135-7* **$24.95**
The Autobiography of Benjamin Franklin has probably been more extensively read than any other American historical work, and no other book of its kind has had such ups and downs of fortune. Franklin lived for many years in England, where he was agent... Biographies/History Pages 332

Name	
Email	
Telephone	
Address	
City, State ZIP	

☐ **Credit Card**   ☐ **Check / Money Order**

Credit Card Number	
Expiration Date	
Signature	

Please Mail to: Book Jungle
PO Box 2226
Champaign, IL 61825
or Fax to: 630-214-0564

ORDERING INFORMATION
web*: www.bookjungle.com*
email*: sales@bookjungle.com*
fax*: 630-214-0564*
mail*: Book Jungle PO Box 2226 Champaign, IL 61825*
or PayPal *to sales@bookjungle.com*

Please contact us for bulk discounts

DIRECT-ORDER TERMS

**20% Discount if You Order
Two or More Books**
Free Domestic Shipping!
Accepted: Master Card, Visa,
Discover, American Express

www.ingramcontent.com/pod-product-compliance
Lightning Source LLC
Chambersburg PA
CBHW080957020726
47505CB00009B/2236